# PRACTICAL PROBLEMS in MATHEMATICS
## EMERGENCY SERVICES

# Delmar's PRACTICAL PROBLEMS in MATHEMATICS

- Practical Problems in Mathematics for Automotive Technicians, 5e
  George Moore
  Revised by Larry and Todd Sformo
  Order # 0-8273-7944-7

- Practical Problems in Mathematics for Carpenters, 6e
  Harry C. Huth
  Order # 0-8273-4579-8

- Practical Problems in Mathematics for Drafting and CAD, 2e
  John C. Larkin
  Order # 0-8273-1670-4

- Practical Problems in Mathematics for Electricians, 5e
  Herman and Garrard
  Order # 0-8273-6708-2

- Practical Problems in Mathematics for Electronics Technicians, 5e
  Herman and Sullivan
  Order # 0-8273-6761-9

- Practical Problems in Mathematics for Graphic Communications, 2e
  Ervin A. Dennis
  Order # 0-8273-2946-3

- Practical Problems in Mathematics for Health Occupations
  Louise M. Simmers
  Order # 0-8273-677

- Practical Problems in Mathematics for Heating and Cooling
  Russell B. DeVore
  Order # 0-8273-7948

- Practical Problems in Mathematics for Industrial Technology
  Donna Boatwright
  Order # 0-8273-6974-

- Practical Problems in Mathematics for Manufacturing, 4e
  Dennis D. Davis
  Order # 0-8273-6710-4

- Practical Problems in Mathematics for Masons, 2e
  John E. Ball
  Order # 0-8273-1283-0

- Practical Problems in Mathematics for Welders, 4e
  Schell and Matlock
  Order # 0-8273-6706-6

## Online Services

**Delmar Online**
To access a wide variety of Delmar products and services on the World Wide Web, point your browser to:
http://www.cengage.com/delmar
or email: delmar.help@cengage.com

**cengage.com**
To access International Cengage Publishing's home site for information on more than 34 publishers and 20,000 products, point your browser to:
http://www.cengage.com
or email: findit@kiosk.cengage.com

# PRACTICAL PROBLEMS in MATHEMATICS EMERGENCY SERVICES

### Thomas B. Sturtevant

*Chattanooga State
Technical Community College
Chattanooga, Tennessee*

Australia • Brazil • Japan • Korea • Mexico • Singapore • Spain • United Kingdom • United States

**Practical Problems In Mathematics For Emergency Services**
Thomas B. Sturtevant

Publisher: Alar Elken

Acquisitions Editor: Gregory L. Clayton

Editorial Assistant: Amy Tucker

Cover Illustration: John Kenefic

Executive Marketing Manager: Maura Theniault

Production Manager: Larry Main

Art/Design Coordinator: Nicole Reamer

© 2000 Delmar, Cengage Learning

ALL RIGHTS RESERVED. No part of this work covered by the co reproduced, transmitted, stored or used in any form or by any m or mechanical, including but not limited to photocopying, recore ing, taping, Web distribution, information networks, or informat systems, except as permitted under Section 107 or 108 of the 197 Act, without the prior written permission of the publisher.

For product information and technology assistance,
Cengage Learning Customer & Sales Support, 1-80

For permission to use material from this text or p submit all requests online at **www.cengage.com/p** Further permissions questions can be email **permissionrequest@cengage.com**

Library of Congress Control Number: 99-42748

ISBN-13: 978-0-7668-0420-3

ISBN-10: 0-7668-0420-8

**Delmar**
Executive Woods
5 Maxwell Drive
Clifton Park, NY 12065
USA

Cengage Learning is a leading provider of customized learning solution around the globe, including Singapore, the United Kingdom, Australia, Japan. Locate your local office at **international.cengage.com/region**

Cengage Learning products are represented in Canada by Nelson Edu

For your lifelong learning solutions, visit **delmar.cengage.com**

Visit our corporate website at **www.cengage.com**

**Notice to the Reader**
Publisher does not warrant or guarantee any of the products described herein or perform any independent analysis in connection with any of the product in herein. Publisher does not assume, and expressly disclaims, any obligation to obtain and include information other than that provided to it by the manufact expressly warned to consider and adopt all safety precautions that might be indicated by the activities described herein and to avoid all potential hazards. B tions contained herein, the reader willingly assumes all risks in connection with such instructions. The publisher makes no representations or warranties of not limited to, the warranties of fitness for particular purpose or merchantability, nor any such representations implied with respect to the material set publisher takes no responsibility with respect to such material. The publisher shall not be liable for any special, consequential, or exemplary damages resulti the readers' use of, or reliance upon, this material.

Printed in the United States of America
3 4 5 6 7     15 14 13 12 11
FD313

# Contents

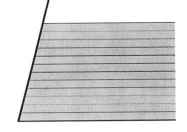

**Preface / vii**

### SECTION 1  BASIC CONCEPTS

Unit 1  Introduction to Numbers / 1
Unit 2  Addition of Positive Integers / 8
Unit 3  Subtraction of Positive Integers / 13
Unit 4  Multiplication of Positive Integers / 17
Unit 5  Division of Positive Integers / 22
Unit 6  Negative Integers, Properties of Zero and One, Exponents and Square Roots / 27
Unit 7  Combined Operations with Integers / 34

### SECTION 2  COMMON FRACTIONS

Unit 8  Introduction to Common Fractions / 39
Unit 9  Addition of Common Fractions / 48
Unit 10  Subtraction of Common Fractions / 54
Unit 11  Multiplication of Common Fractions / 59
Unit 12  Division of Common Fractions / 64
Unit 13  Combined Operations with Common Fractions / 67

### SECTION 3  DECIMAL FRACTIONS

Unit 14  Introduction to Decimal Fractions / 73
Unit 15  Addition of Decimal Fractions / 77
Unit 16  Subtraction of Decimal Fractions / 83
Unit 17  Multiplication of Decimal Fractions / 86
Unit 18  Division of Decimal Fractions / 92
Unit 19  Decimal and Common Fraction Equivalents / 95
Unit 20  Combined Operations with Decimal Fractions / 98

### SECTION 4  PERCENT, INTEREST, AVERAGES, AND ESTIMATES

Unit 21  Percent and Percentages / 101
Unit 22  Interest and Discounts / 109
Unit 23  Averages and Estimates / 115

## SECTION 5  MEASUREMENT

Unit 24  Introduction to Measurement  /  119
Unit 25  Length Measurements  /  123
Unit 26  Area and Pressure Measurements  /  129
Unit 27  Solid and Fluid Volume Measurements  /  133
Unit 28  Mass and Density Measurements  /  138
Unit 29  Temperature Measurements  /  141

## SECTION 6  STATISTICS, CHARTS, AND GRAPHS

Unit 30  Introduction to Statistics  /  145
Unit 31  Line Graphs  /  151
Unit 32  Bar Graphs  /  157
Unit 33  Pie Charts  /  164

## SECTION 7  FORMULAS AND EQUATIONS

Unit 34  Introduction to Formulas and Equations  /  171
Unit 35  Common Emergency Service Formulas and Equations  /  173

## APPENDIX

SECTION 1  English Relationships  /  181
SECTION 2  Metric Relationships  /  182
SECTION 3  English-Metric Equivalents  /  184
SECTION 4  Fraction/Decimal Equivalents  /  185
SECTION 5  Powers and Roots of Numbers  /  186
SECTION 6  Common Emergency Service Abbreviations  /  187

## GLOSSARY  /  189

## ANSWERS TO ODD-NUMBERED PROBLEMS  /  192

# *Preface*

*The American Heritage Dictionary of the English Language, Third Edition*, defines *mathematics* as "the study of the measurement, properties, and relationships of quantities, using numbers and symbols." Mathematical operations are commonly used in most aspects of emergency services. Maintaining equipment inventory, developing and maintaining department budgets, performing calculations on the emergency scene, and reporting emergency service statistical data are a few examples of activities requiring a fundamental knowledge of mathematics. Every member in the emergency services should possess and maintain at least a basic understanding of mathematical principles.

*Practical Problems in Mathematics for Emergency Services* was designed to provide a basic review of math skills and problem-solving skills for personnel in fire, emergency medical, hazardous material, rescue, and other branches of the emergency service field. A wide variety of individuals may find this book useful. For example, this text might be helpful to individuals reviewing math topics and problems typically found on civil service examinations, state or national certification examinations, and promotional tests. In addition, students engaged in the study of emergency service–related subjects in a college course, department training, state training, academy program, or self-study may find this text helpful. Finally, individuals who desire increased knowledge, skill, and retention of mathematics and its application to the emergency service field will find this text useful.

The text is organized into units, each focusing on specific math concepts. Each unit contains a concise statement of objectives. In addition, several solved examples are provided followed by a variety of practical problems. Each of the problems relate to current aspects of the emergency service field. Units are grouped in sections containing several related objectives. An Instructor's Guide is available that includes answers to all problems as well as two achievement reviews.

The glossary includes definitions of both mathematical and technical terms. The appendix provides additional math-related reference material for the emergency service field. Finally, answers to odd-numbered problems are provided.

## DEDICATION

This book is dedicated to my former supervisors, Rick, Ed, and Charlie and to Dr. Dan Quarles. The extra time, constant encouragement, and continued support have not gone unnoticed or unappreciated. Thank you for all you have done and for your continued friendship over the years.

## ACKNOWLEDGMENTS

First, the author would like to thank his wife, Karen Sturtevant, who reviewed the entire manuscript and solved all the problems. Her insight and suggestions provided invaluable support and constant encouragement. Second, the author would like to thank all the great folks at Delmar, with special thanks to Julie. It was a pleasure working with such positive and supportive folks.

## REVIEWERS

The author and publisher wish to thank the following individuals who reviewed the manuscript and provided helpful suggestions:

William Cheaqui
Kansas City Fire Academy
Kansas City, MO

David Hauger
Westmoreland County Community College
New Stanton, PA

James Ferguson
Chemeketa Community College
Salem, OR

Tim Flannery
John Jay College of Criminal Justice
North Brunswick, NJ

Thomas Feierabend
Mt. San Antonio College
Walnut, CA

Keith Heckler
Bowie, MD

# Basic Concepts

## SECTION 1

### Unit 1  INTRODUCTION TO NUMBERS

**OBJECTIVES**

Upon completion of this unit, the student should be able to

- identify the types of numbers commonly used in emergency services.
- briefly explain the base 10 numbering system.
- write numbers in words and in standard form.
- round numbers to a given place value.

**NUMBERS**

*Arithmetic* comes from two Greek words, "arithmos" meaning number and "techne" meaning art or skill. Arithmetic is the study of numbers and their use in the basic operations of addition, subtraction, multiplication, and division. The most common numbers used in emergency services are *real numbers* which include whole numbers (0, 1, 2, . . .,) integers, fractions, and decimal fractions. *Integers* are positive and negative whole numbers with no fractional or decimal parts ( . . . –3, –2, –1, 0, 1, 2, 3, . . .). These represent whole units of something. *Positive integers*, also referred to as counting numbers, are the numbers 1, 2, 3, 4, and so on. *Negative integers* are the numbers –1, –2, –3, –4, and so on. The integer 0 is neither positive nor negative. Integers can be thought of as points on a line extending right in a positive and left in a negative direction.

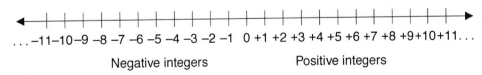

Fractions and decimal fractions, also referred to as *rational numbers*, are numbers used to express increments between integers. Fraction and decimal fraction numbers will be discussed in later units.

**NUMBERING SYSTEM**

Different numbering systems have been developed through history. The most common numbering system—the one we will be using—is called *base 10* or the *decimal system*. In the base 10 system, objects are counted in groups of ten using *digits*, (0, 1, 2, 3, 4, 5, 6, 7, 8, and 9). Digits are grouped in three's, called *periods*, and are separated by a comma. For example,

one period         423
two periods      365,298
three periods     945,462,432

Each digit has its own place value determined by the location within the number. When using positive integers, the digit in a number farthest to the right has a place value known as *ones*. The following figure shows the names and groupings of the first twelve place values. Period names are in capital letters.

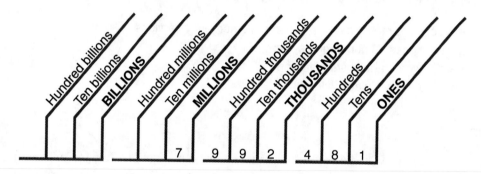

In the number 7,992,481, the digit 7, for example, has a place value of millions while the digit 8 has a place value of tens.

## WRITING NUMBERS

Emergency service personnel may be called upon to write numbers in both standard form and in words. The *standard form* uses digits when writing numbers. To write a number in the standard form, write each number using digits, being sure to separate periods (groups) with a comma. The number 31,425,687 is in standard form and has three periods. To write a number in words, name the number in each place value starting from the left. In the location of the comma, write the period name. The same number written in words is: thirty-one million four hundred twenty-five thousand six hundred eighty-seven.

**Examples:** *Standard form*               *Word form*
               24-hour shift                 twenty-four hour shift
               occupant load of 402      occupant load of four hundred two
               1,500 ml of normal saline    one thousand five hundred ml of normal saline
               682,029 square feet        six hundred eighty-two thousand twenty-nine square feet
               1,004,300 dollars           one million four thousand three hundred dollars

## NUMBERS AND UNITS

In the preceding examples, numbers were associated with units (for example, a thing, size, or quantity). During arithmetic operations it is important to keep the assigned units straight. It makes no sense to add 8 gloves to 4 pieces of apparatus. *Like units* should be used in arithmetic operations. This is especially true when using formulas, in which like units can be combined through addition or subtraction and canceled. In addition, units can be expressed in a number of ways. Consider the following examples:

*gallons per minute* is the same as gpm and gal/min
*pounds per square inch* is the same as psi and lb/in$^2$

Additional discussion of units will be provided later in the text.

## ROUNDING NUMBERS

During certain calculations in the emergency services, especially those made on the emergency scene and when presenting statistical data, an approximation of a number rather than the exact number is used. The process of giving an approximate value instead of an exact number is called *rounding*. Numbers are rounded to a specified place value. To round a number to a specified place value, look at the first digit to the right of the specified place value. If the digit is less than 5, that digit and the remaining digits to the right are replaced with zeros. The number 9,429 rounded to the nearest hundreds is 9,400. Another example follows.

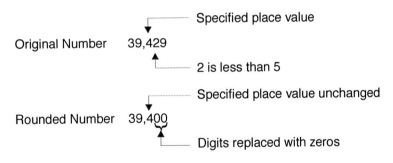

If the digit to the right of the specified digit is greater than or equal to 5, the digit in the specified place value is increased by one, and the remaining digits to the right are replaced with zeros. The whole number 17,681 rounded to the nearest hundreds is 17,700.

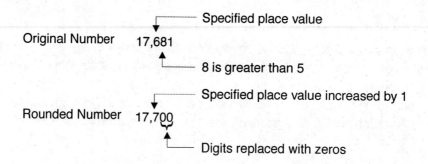

Some additional examples of rounding numbers follow.

**Example 1:** Round 65,736 to the nearest thousand.

**Solution:**

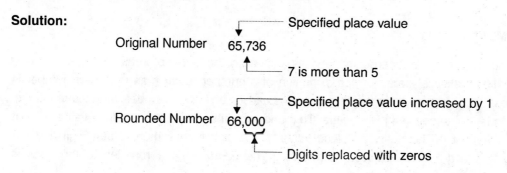

**Example 2:** Round 1,350,256 to the nearest million.

**Solution:**

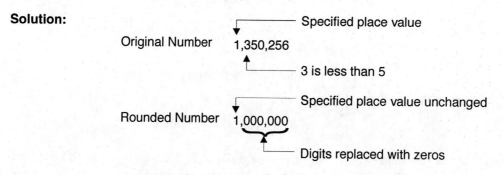

**Example 3:** Round 52,958 fire ground injuries to the nearest hundred.

**Solution:**

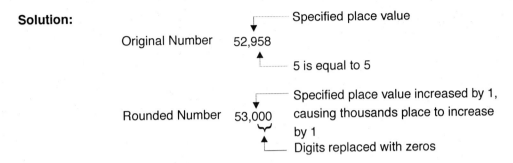

 **CALCULATORS**

Calculators are a tool commonly used in education, the emergency services, and in everyday life to speed the process of mathematical operations. In emergency services, they are used to calculate fire loads, drug dose, pressures, supply inventory, and budgets. Several types and brands of calculators are available providing a wide range of capabilities. The calculator shown at the end of this paragraph is a special calculator designed for hydraulic calculations. It can be used for typical math operations as well. Not all calculators operate in the same manner. For example, the order in which numbers and operations are entered can be different from one calculator to another. It is important to become familiar with your calculator. First, read the instructions, and then practice by performing simple math problems until you become comfortable with its operation. Selected sections in this text include an introduction to the use of an algebraic logic calculator.

## PRACTICAL PROBLEMS

Write the following numbers in words.

1. 356 _____     3. 26,498 _____

2. occupant load of 523 _____     4. 6,375,314 dollars _____

Write the following numbers in standard form:

5. three hundred fourteen medical runs _____

6. six hundred eight _____

7. one hundred four thousand six square feet _____

8. three million two hundred seventy-four _____

Round the following numbers to the given place value.

9. 243 (tens) _____

10. 648 psi (tens) _____

11. 2,368 (hundreds) _____

12. 5,839 residential fires (hundreds) _____

13. The 1996 November/December issue of the *NFPA Journal* reports that during 1995, a total of 5,230 firefighters in the United States were injured while responding to or returning from an incident. In a summary report to your chief, you write, "During 1995, approximately _____ thousand firefighters were injured while responding to or returning from an incident." (Hint: Round 5,230 to the thousands place value.) _____

14. The recently approved 1998 fiscal budget for the ABC Fire Department is $16,354,453. The local paper reports that the fire department's approved budget is $_____ million. _____

15. During a recent residential fire, the ABC Fire Department pumped a total of 16,450 gallons of water from three hand lines and one exposure line. The incident fire report indicates that approximately _____ thousand gallons of water were used during the incident.

_____

# Unit 2  ADDITION OF POSITIVE INTEGERS

**OBJECTIVE**

Upon completion of this unit, the student should be able to

- add positive integers.

**BASIC PRINCIPLES OF ADDITION**

The mathematical process for finding the total value of two or more numbers is called *addition*. The basic symbol for addition is the plus (+) sign. Several key phrases used to indicate the need for addition include *add*, *plus*, *sum*, and *total*. Each number being added is called an *addend*. The equal symbol (=) is used to indicate the *sum* of the numbers. The first step in adding positive integers is to write the numbers, in standard form in a column, being sure to place digits in their appropriate place value. Next, add each column beginning with the ones column and moving left. Write the total of each place value under each column. For example, the sum of 23 + 25 is 48.

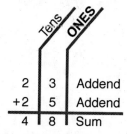

When the sum of digits in a column is greater than 9, the addition process requires *carrying*. The addition of 5 + 49 requires carrying to obtain the sum of 54.

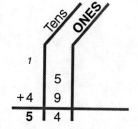

First add the ones column. 5 + 9 = 14 (1 ten and 4 ones), write the 4 in the ones column, and carry the 1 to the tens column.

Finish by adding the tens column. 1 + 4 = 5. The sum of 5 + 49 = 54.

Unit 2  ADDITION OF POSITIVE INTEGERS  9

**Example 1:** Add: 82 + 15.

**Solution:**
```
  82
 +15
 ---
  97
```

**Example 2:** During 1997, A shift installed 287 smoke detectors, B shift installed 354 smoke detectors, and C shift installed 174 smoke detectors. What was the total number of smoke detectors installed for 1997? _____ .

**Solution:**
```
   21
  287
  354
 +174
 ----
  815
```

 **CALCULATOR USE**

The addition key on most calculators is marked with the plus sign (+). Most calculators use algebraic logic. Enter the first number, press the (+) key, enter the next number, press the (+) key, and continue until the last number is entered. The equal sign (=) can be pressed to obtain the total.

**PRACTICAL PROBLEMS**

1. 29 + 14 _____

2. 65 + 42 _____

3. 150 + 50 _____

4. 100 + 25 + 40 _____

5. 1,204 + 34,963 + 64 _____

6. Upon arrival at a structure fire, Engine 1 dropped 150 feet of 2½-inch hose, Engine 3 dropped 600 feet of 3-inch hose, and Engine 5 dropped 750 feet of 4-inch hose. How much hose is on the ground at this incident? _____

10   Section 1   Basic Concepts

7. According to the following illustration, what is the total gpm flow for the pumper? _____

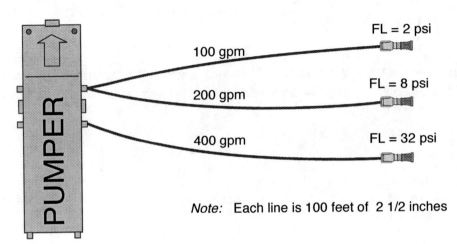

Note:  Each line is 100 feet of 2 1/2 inches

8. The new 1999 fire department budget distributes funds as follows:

| Operations | $6,365,215 |
| Prevention | $ 750,240 |
| Maintenance | $1,320,226 |
| Training | $ 500,649 |

What is the total fire department 1999 budget? _____

9. The total stopping distance for an apparatus is equal to the perception distance, reaction distance, and braking distance. According to this illustration, the total stopping distance for this apparatus is _____.   _____

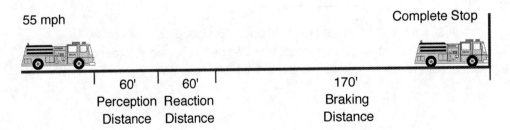

Unit 2 ADDITION OF POSITIVE INTEGERS 11

10. Complete the following 1998 first quarter run report.

*First Quarter Run Report, 1998*

|        | January | February | March | Totals |
|-------:|:-------:|:--------:|:-----:|:------:|
| Fire   | 10      | 16       | 7     |        |
| EMS    | 22      | 31       | 19    |        |
| Rescue | 3       | 1        | 4     | 8      |
| Hazmat | 2       | 5        | 2     |        |
| Total  | 37      |          |       | 122    |

11. A pump operator inventories the number of nozzles on the apparatus. She finds 2 automatic nozzles (1½-inch), 4 automatic nozzles (2½ inch), 2 fixed gpm nozzles, and 4 forestry nozzles (1-inch). What is the total number of nozzles on the apparatus? _____

12. The EMS director for a county compiles the following three-month graph showing the number of medical runs for the north and south side of the county. What is the north side's, south side's, and combined total number of medical runs for the three-month period?

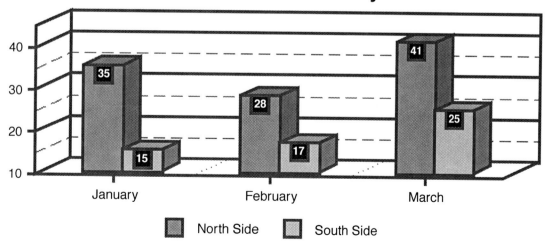

North side:
South side:
Combined:

13. The apparatus for a volunteer fire department maintains the following on-board water capacity:

    Engine 1:   500 gallons          Engine 2:   300 gallons
    Tanker 1:   1,000 gallons        Brush 1:    750 gallons

    What is the total on-board water capacity for the volunteer fire department? _____

14. A regional hazardous material team maintains the following chemical suits:

    |               | Level A | Level B |
    |---------------|---------|---------|
    | Hazmat Van    | 2       | 4       |
    | Hazmat Trailer| 4       | 4       |
    | Supply        | 5       | 10      |

    What is the total number of chemical suits maintained by the hazmat team? _____

15. The ABC Fire Department requires that all members receive 8 hours of medical training, 16 hours of hazmat training, 16 hours of tactics and strategy training, 4 hours of incident command training, and 8 hours of safety training. What is the total number of training hours required for each member? _____

# Unit 3 SUBTRACTION OF POSITIVE INTEGERS

## OBJECTIVE

Upon completion of this unit, the student should be able to

- subtract positive integers.

## BASIC PRINCIPLES OF SUBTRACTION

The mathematical process for finding the difference between two numbers is called *subtraction*. The basic symbol for subtraction is the minus (–) sign. Several key phrases used to indicate the need for subtraction include *difference*, *minus*, *subtracted from*, *less than*, and *remaining*. The number being subtracted from is called the *minuend*. The number to be subtracted is called the *subtrahend*. The equal symbol (=) is used to indicate the difference. As with addition, the first step in the subtraction process is to write the numbers in a column, being sure to place digits in their appropriate place value. Next, subtract digits in each column beginning with the ones column and moving left. Write the difference of each place value under each column. For example, the difference between 75 and 32 is 43.

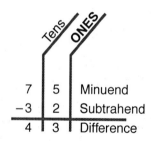

```
  Tens  ONES
   7  |  5    Minuend
  -3  |  2    Subtrahend
  ─────────
   4  |  3    Difference
```

When a digit in the subtrahend is larger than the one above it in the minuend, the subtraction process requires *borrowing*. The subtraction of 74 – 58 requires borrowing to obtain the difference of 16.

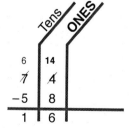

| 6 + 1 | | | | 6 +(1) 10 | | | | 6 | 14 |
|---|---|---|---|---|---|---|---|---|---|
| 7 | 4 | Minuend | | 7̸ | 4 | | | 7̸ | 4̸ |
| –5 | 8 | Subtrahend | | –5 | 8 | | | –5 | 8 |
| | | Difference | | | | | | 1 | 6 |

Because 8 is greater than 4, borrowing is required. 7 can be rewritten as 6 tens + 1 ten.

Borrow 1 ten and place it in the ones column.

Add the borrowed 10 to 4 and subtract the digits in each column.

*13*

14  Section 1  Basic Concepts

To check for accuracy of subtraction, add the subtrahend and the difference. The answer should equal the minuend.

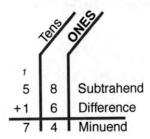

**Example 1:** Subtract: 537 − 225.

**Solution:**
```
 537
−225
 312
```

**Example 2:** At the beginning of a six-month period, a total of 644 boxes of 4-inch by 4-inch gauze pads were stored in supply. During the period, 372 boxes were used. How many boxes remain?

**Solution:**
```
 5 14
 644
−372
 272
```

 **CALCULATOR USE**

The subtraction key on most calculators is marked with a minus sign (−). For calculators that use algebraic logic, enter the first number, press the (−) key, enter the number to be subtracted, then press the equal (=) key.

**PRACTICAL PROBLEMS**

1. 19 − 5  _____
2. 64 − 42  _____
3. 375 − 67  _____
4. 784 − 392  _____

5. 472 – 274  _____

6. A pumper operating on the fire ground is discharging 625 gpm through two 2½-inch lines and one 1¾-inch line. If one of the 2½-inch lines, flowing 250 gpm, is shut down, how many gpms is the pumper still discharging?  _____

7. During an initial medical assessment, a patient's pulse rate is 122 beats per minute (bpm). After 30 minutes, the pulse rate drops by 15 bpm. What is the patient's pulse rate?  _____

8. A fork lift accident punctures a 500-gallon tank spilling hydrazine onto the floor and into the storm drain. After the hazmat team stops the leak, only 325 gallons of hydrazine remain in the tank. Assuming the tank was full when the accident occurred, how many gallons of hydrazine were spilled?  _____

9. A fire prevention's annual budget summary indicates both budgeted expenses and actual expenses. What is the difference between budgeted and actual expenses in each category, as well as the overall difference between budgeted and actual expenses?

**Fire Prevention
Annual Budget Summary**

|  | Budgeted | Actual | Difference |
|---|---|---|---|
| Personnel (Salaries and Benefits) | 175,842 | 187,681 |  |
| Operations (Operating/Maintenance Supplies) | 32,000 | 34,295 |  |
| Equipment (Capital) | 2,300 | 1,784 |  |
| Total |  |  |  |

Personnel:  _____
Operations:  _____
Equipment:  _____
Total:  _____

10. If a pump's discharge pressure for an attack line is 178 psi and the total friction loss for the hose line 79 psi, what is the remaining psi at the nozzle? _____

11. The fire prevention bureau is responsible for conducting 53 fire safety inspections per quarter. Halfway through the first quarter the bureau has completed 34 inspections. How many inspections are left in the quarter? _____

12. At the start of a specialized diet/fitness program, a firefighter weighs 240 pounds. After four months on the program, the firefighter weighs 198 pounds. How much weight has the firefighter lost? _____

13. A city fire department has a total of 483 sections of hose in service. During annual hydrostatic testing of hose, 29 sections are removed from service. How many sections of hose are still in service? _____

14. Benzene has a flash point of 12°F while isopropyl alcohol has a flash point of 53°F. What is the difference between the flash points? _____

15. The apparatus maintenance department of a fire department requires that all vehicles receive a thorough inspection at 60,000 miles. If a vehicle has 49,765 miles, how many more miles can it be driven before the inspection is required? _____

# Unit 4  MULTIPLICATION OF POSITIVE INTEGERS

**OBJECTIVE**

Upon completion of this unit, the student should be able to

- multiply positive integers.

**BASIC PRINCIPLES OF MULTIPLICATION**

*Multiplication* is the shortened process for repeated addition of a number. The basic symbol for multiplication is the times (×) sign. Other symbols used for multiplication include a dot placed between two numbers and placing numbers side by side separated by parentheses. Each of the following indicates that multiplication is required:

$$3 \times 5, 3 \cdot 5, \text{ and } (3)(5).$$

Several key phrases used to indicate the need for multiplication include *times*, *multiply*, and *product*. The number being multiplied is called the *multiplicand*, and the number that indicates how many times to (in effect) add the multiplicand is the *multiplier*. The multiplicand and the multiplier are also called *factors*. The answer in multiplication is called the *product*. When 6 is multiplied by 3, the product is 18. The same result is achieved when 6 is added three times (6 + 6 + 6 = 18). Note that *carrying* (bringing a digit to the next higher place value) occurs when the product is larger than 9.

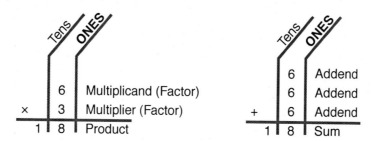

The preceding example is a single-digit multiplication problem. Single-digit multiplications through the 9's should be memorized. Multiple-digit multiplication requires the repeated use of single-digit multiplication. First write the numbers in a column with the larger number, as the multiplicand, on top and with the multiplier, the smaller number, below. The results are the same regardless of order: 2 × 4 is the same as 4 × 2. However, multiple-digit multiplication is easier when the larger number is the multiplicand. Each number in the multiplicand is multiplied by each number in the multiplier. During this process several results are obtained called *partial products*. After each multiplicand digit is multiplied by

18  Section 1  Basic Concepts

each multiplier digit, the partial products are added for the *final product*. For example, the multiplication of 12 × 36 requires repeated use of single-digit multiplication.

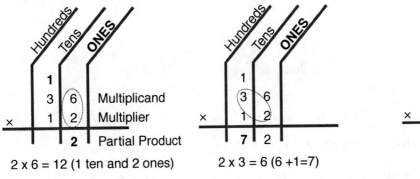
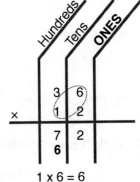

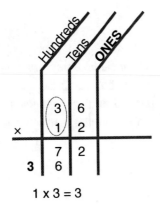

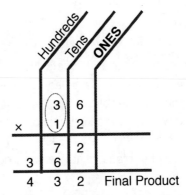

Last step is to add the partial products, remembering to carry when the sum is greater than 9.

**Example 1:** Find the product: 5 × 12.

**Solution:**
```
  1
 12
 ×5
 ──
 60
```

**Example 2:** The standard length of a section of hose in the fire service is 50 feet. If a pumper lays 14 sections of 2½-inch hose, how long is the hose lay?

**Solution:**  50 feet per ~~section~~ of 2½-inch hose
×14 ~~sections~~ of 2½-inch hose
700 feet of 2½-inch hose (Note that the unit "section" cancels out.)

### CALCULATOR USE

The multiplication key on most calculators is marked with the multiplication sign (×). For calculators that use algebraic logic, enter the first factor, press the (×) key, enter the next factor, and then press the equal (=) key to display the product. If more than two factors are multiplied, you may enter all the factors with the multiplication key pressed between each factor, until all the factors have been entered. The final step, then, is to press the (=) key.

### PRACTICAL PROBLEMS

1. Complete the multiplication table that follows.

|   | 0 | 1 | 2 | 3 | 4 | 5 | 6 | 7 | 8 | 9 |
|---|---|---|---|---|---|---|---|---|---|---|
| 1 |   |   |   |   |   |   |   |   |   |   |
| 2 |   |   |   |   |   |   |   | 14 |   |   |
| 3 |   |   |   |   |   |   |   |   |   |   |
| 4 |   |   |   | 12 |   |   |   |   |   |   |
| 5 |   |   |   |   |   |   |   |   |   |   |
| 6 |   |   |   |   |   |   |   |   |   |   |
| 7 |   |   |   |   |   |   |   |   |   |   |
| 8 |   |   |   |   |   | 40 |   |   |   |   |
| 9 |   |   |   |   |   |   |   |   |   |   |

2. 4 × 37  _____

3. 29 × 53  _____

4. 63 × 446  _____

5. 385 × 668  _____

20   Section 1   Basic Concepts

6. During a recent structural fire, Engine 5 maintained a pump discharge of 275 gpm for 14 minutes. What was the total gpm discharge for Engine 5 over the 14-minute period?   _____

7. An ambulance company responds to an average of 3 calls per day. What is the average number of calls over a 30-day period?   _____

8. Water weighs approximately 8 pounds per gallon. What is the approximate weight of 50 gallons of water?   _____

9. A patient's pulse rate is 68 beats per minute. How many times will the heart beat in 9 minutes?   _____

10. The ABC Hotel has 10 exit doors with an exit capacity of 32 for each door. What is the total exit capacity of all doors combined?   _____

11. An area 12 inches by 12 inches is divided into 1-inch by 1-inch sections as shown in the following diagram. How many 1-inch by 1-inch sections are in the 12-inch by 12-inch area?   _____

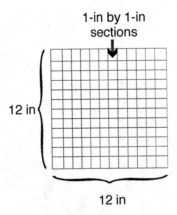

12. Part of an annual pump test requires a 1,250 gallons per minute (gpm) pump to discharge 625 gpm for 10 minutes. How many gallons does the pump discharge during this period?   _____

13. A paramedic examines the following six-second strip of an electrocardiogram (EKG). How many times per minute is the patient's heart beating? (Hint: There are 10 six-second periods in one minute.)   _____

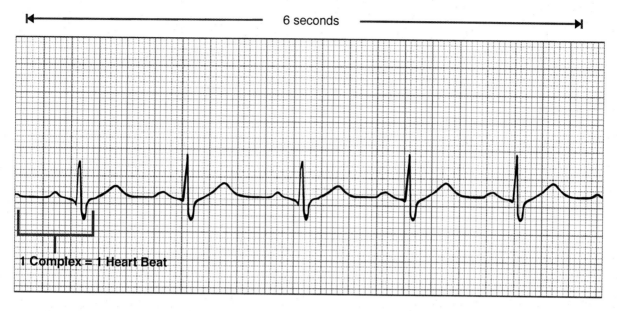

14. When emergency service personnel work a 24/48 (24 hours on and 48 hours off) shift schedule, they are on-duty approximately 10 days per month. Approximately how many days are emergency service personnel on duty in a year?

15. The heart pumps about 65 milliliters (ml) of blood every time it beats. If a heart is beating 65 times per minute, how many ml of blood does the heart pump per minute and per hour?

    Per minute:
    Per hour:

# Unit 5  DIVISION OF POSITIVE INTEGERS

## OBJECTIVE

Upon completion of this unit, the student should be able to

- divide positive integers.

## BASIC PRINCIPLES OF DIVISION

*Division* is the shortened process for repeated subtraction of a number. The process can be viewed as a way to determine how many times one number is contained in another or how to separate a quantity of things into groups of equal numbers. The basic symbols for division are signs (÷) and (/) and the division bracket ( $\overline{)}$ ). Each of the following indicates that division is required: 5 ÷ 3, 5/3, 3$\overline{)5}$ . Several key phrases used to indicate the need for division include *divided by*, *average*, *per*, *out of*, and *each*. The number being divided is called the *dividend*, and the number that indicates how many times to divide the dividend is called the *divisor*. The answer is the *quotient* and indicates the number of times the divisor is contained in the dividend. When the divisor cannot be contained in the dividend a whole number of times, the number left over is called the *remainder*. When 8 is divided by 2, the quotient is 4. In other words, 2 goes into 8, 4 times. The same results are achieved when 2 is repeatedly subtracted from 8.

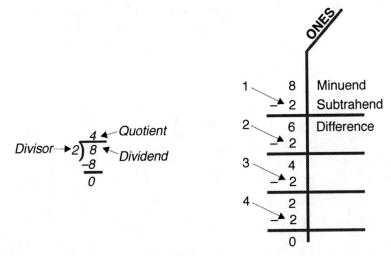

The preceding example is a single-digit division problem. Multiple-digit division can require several steps. The process begins by determining the number of times the divisor can be divided into the first digit or digits of the dividend, starting from the left. This number is placed in the quotient directly above the last digit used in the dividend. Next, multiply the quotient by the divisor and place the product under the digits just used in the dividend and subtract the two numbers. The next digit in the dividend is placed by the new number obtained from subtracting. The process is repeated until all the digits in the dividend have been used. The division problem, 3,767 ÷ 4, requires several steps, as indicated in the following diagrams.

```
              9   Quotient              94                    941
Divisor   4) 3767  Dividend         4) 3767               4) 3767
             36                        36                    36
             ──                        ──                    ──
             16                        16                    16
                                       16                    16
                                       ──                    ──
                                       07                    07
                                                              4
                                                              ─
                                                              3
```

Because 4 is larger than the first digit 3, 37 is used; 4 goes into 37 a total of 9 times (4 x 9 = 36), 37 − 36 is 1; 6 is the next digit in the dividend and is placed by the 1.

4 goes into 16 a total of 4 times (4 x 4 = 16), 16 − 16 = 0; 7 is the next digit in the dividend and is placed by the 0.

4 goes into 7 only 1 time (4 x 1 = 4), 7 − 4 = 3; no other numbers are left in the dividend so 3 is the remainder.

To check your answer, multiply the quotient by the divisor and add the remainder if applicable. The result should be the dividend. Hence, in the preceding example: 941 x 4 = 3764; 3764 + 3 = 3767.

**Example 1:** Divide 4,824 by 18.

**Solution:**
```
            268
       18) 4824
            36
            ──
            122
            108
            ───
            144
            144
            ───
              0
```

**Example 2:** According to the National Fire Protection Association (NFPA), approximately 4,035 civilians lost their lives in home fires during 1996 (*NFPA Journal*, September/October 1997). Approximately how many lives were lost each month during 1996?

24  Section 1  Basic Concepts

**Solution:**
```
      336
   12)4035
      36
      ‾‾
       43
       36
       ‾‾
        75
        72
        ‾‾
         3 (remainder)
```

Approximately 336 lives were lost each month.

### CALCULATOR USE

The division key on most calculators is marked with a (÷) sign. For calculators that use algebraic logic, enter the dividend, then press the (÷) key, then enter the divisor, and press the (=) key. The quotient will be displayed. If a decimal point appears, followed by other digits (as in 69.35), the division involved a remainder or did not divide "evenly."

### PRACTICAL PROBLEMS

1. 465 ÷ 7 _____

2. 9,585 ÷ 45 _____

3. 716,821 ÷ 4 _____

4. 268 ÷ 14 _____

5. 12,032 ÷ 376 _____

6. An EMS director orders 12 new stethoscopes for 324 dollars. How much was paid for each stethoscope? _____

7. Engine 26 has a total complement of 2,100 feet of hose. How many sections of hose are on this apparatus (if each section of hose is 50 feet in length)? _____

8. A fire prevention bureau is responsible for inspecting 288 commercial buildings each year. How many buildings must be inspected each month to distribute the workload evenly over the year? _____

9. Engine 5 has an on-board water supply of 1,500 gallons. If one 1¾-inch preconnect is flowing 125 gpm, how long will the on-board water supply last? (Hint: When you divide 1,500 gallons by 125 gpm, or gallons per minute, the gallons unit cancels, leaving minutes as the final unit.) _____

10. The quarterly run report for a county's three medic units follows. What is the average number of runs per quarter for each medic unit? (Hint: The average for a group of values is determined by adding the values and then dividing by the number of values.)

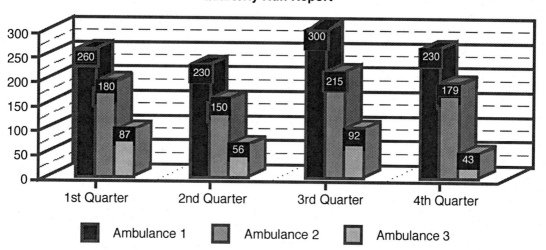

**Quarterly Run Report**

Ambulance 1: _____
Ambulance 2: _____
Ambulance 3: _____

11. During a recent fire, Engine 3 pumped a total of 60,000 gallons over a two-hour period. How many gallons were pumped each minute (gallons per minute, or gpm)? (Hint: "Per" indicates divide the first unit by the second unit.) _____

12. During the size-up of a fire, it is estimated that approximately 625 gpm are needed on the fireground. If three pumpers are on-scene, how many gpm must each pumper provide to evenly distribute the total gpm? _____

13. If a force of 625 pounds is applied to an area of 144 square inches, what is the pressure in pounds per square inch? (Hint: The unit of pounds per square inches can be indicated as psi or lbs/in$^2$. Note that when pounds is divided by square inches, the unit becomes pounds per square inch.) _____

26  Section 1  Basic Concepts

14. One way to estimate the needed flow (gpm) for a structure is to divide the volume (ft$^3$) of the structure by the constant 100 ft$^3$/gpm. The formula, called the Iowa State Formula, is expressed as $NF = V/100$ ft$^3$/gpm. If the volume of a structure is 21,600 ft$^3$, what is the needed flow (gpm)? _____

15. The National Fire Academy Formula for determining needed flow divides the area in square feet by the constant 3 ft$^2$/gpm. If the area of a structure is 1,200 ft$^2$, what is the needed flow (gpm)? _____

# Unit 6 NEGATIVE INTEGERS, PROPERTIES OF ZERO AND ONE, EXPONENTS AND SQUARE ROOTS

## OBJECTIVES

Upon completion of this unit, the student should be able to

- perform arithmetic operations using negative integers.
- explain the special considerations for zero and one in arithmetic operations.
- explain the concepts of exponents and square roots.

## NEGATIVE INTEGERS

As mentioned previously, integers are the numbers –3, –2, –1, 0, 1, 2, 3, and so on. *Negative integers* are simply the numbers –1, –2, –3, –4, and so on. Compared to positive integers, negative integers extend to the left on the number line, as shown in the following diagram. A number with no sign or a plus (+) sign before it is positive, while a number with a negative (–) sign before it is negative. The term *signed operations* indicates the use of positive and negative numbers during arithmetic operations. When positive numbers and negative numbers are used in arithmetic problems, they are often contained in parentheses to avoid confusing the operation sign with the arithmetic sign. For example, when –4 is subtracted from +3, the problem is written as (+3) – (–4) or 3 – (– 4).

The *absolute value* of a number is often needed when conducting arithmetic operations with signed numbers. The *absolute value* of a number is the value of the number with no regard for the sign and is indicated with / /. For example, the absolute value of –5 is denoted as /–5/ and is equal to 5. The absolute value of –14 is written as /–14/ = 14.

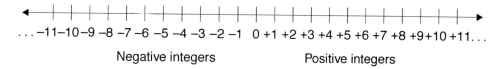

### Addition of Signed Numbers

When two numbers being added have the *same* or *like sign*, the absolute value of the numbers are added and the sign of the original numbers is attached to the sum. For example, the problem (–2) + (–5) is solved by adding their absolute values, I–2I = 2 and I–5I = 5, and attaching the negative sign to the sum. Thus, (–2) + (–5) = I–2I + I–5I = 2 + 5 = 7. The addition problems (+2) + (+5) = +7 and (–2) + (–5) = –7 are illustrated on the number lines that follow.

27

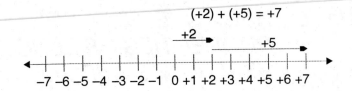

$(+2) + (+5) = +7$

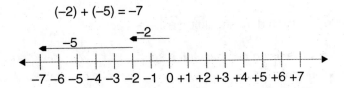

$(-2) + (-5) = -7$

When two numbers being added have *different* or *unlike signs*, the difference of the absolute value of the numbers is found by subtracting the smaller absolute value from the larger. The sign of the number of the larger absolute value is attached to the difference. For example, the problem $(+2) + (-5)$ is solved by subtracting the absolute values and attaching the sign of the number of larger absolute value to the sum. Thus, $(+2) + (-5) = -(|-5| - |+2|) = -(5 - 2) = -3$. The addition problems $(+2) + (-5) = -3$ and $(-2) + (+5) = +3$ are illustrated on the number lines that follow.

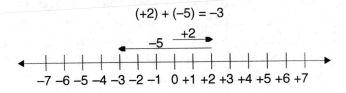

$(+2) + (-5) = -3$

$(-2) + (+5) = +3$

**Subtraction of Signed Numbers**

The subtraction of signed numbers requires the rephrasing of the subtraction problem to an addition problem and following the rules for addition of signed numbers. Rephrasing is accomplished by changing the operation sign from a minus to a plus and switching the sign of the subtrahend (the number being subtracted). Several examples of the process are illustrated here.

Unit 6 NEGATIVE INTEGERS, PROPERTIES OF ZERO AND ONE, EXPONENTS AND SQUARE ROOTS

**Examples:** (+6) − (−4)  is rephrased as  (+6) + (+4) =  (+10)
(−6) − (−4)  is rephrased as  (−6) + (+4) =  (−2)  } (from rules for adding numbers
(−6) − (+4)  is rephrased as  (−6) + (−4) =  (−10) }  of like and unlike signs)

Any subtraction problem can be rephrased as an addition problem.

**Multiplication and Division of Signed Numbers**

When the signs of numbers are the same in a multiplication or division problem, the answer is always positive. When the signs of numbers used are different, the answer is always negative. Several examples follow.

**Examples:** Same Sign          Different Sign
(−3) × (−4) = (+12)     (−3) × (+4) = (−12)
(−20) ÷ (−5) = (+4)     (+20) ÷ (−5) = (−4)

## PROPERTIES OF ZERO AND ONE

The properties of zero and one in arithmetic operations should be memorized.

**Properties of Zero**

Adding zero to a number does not change the number. For example:

25 + 0 = 25
0 + (−14) = (−14)

The same is true when 0 is subtracted from a number. For example: 2,000 − 0 = 2,000. However, when a number is subtracted from zero, the result of rephrasing changes the sign of the number. For example, 0 − 8 = (−8) because the problem is rephrased as 0 + (−8), and the sign of the number of larger absolute value is attached to the sum. When a multiplication problem includes a zero, the product is zero. For example, 0 × 5 = 0 and 5 × 0 = 0. Or 3 × 0 × −20 = 0. Dividing a number by zero is an *undefined* operation.

**Properties of One**

When a number is multiplied or divided by +1, the number does not change. When a number is multiplied or divided by −1, the sign of the number changes. Several examples follow.

**Examples:** Operations with +1
$$1 \times 58 = 58$$
$$(-32) \times 1 = (-32)$$
$$(-22) \div 1 = (-22)$$
$$458 \div 1 = 458$$

Operations with −1
$$-1 \times 58 = -58$$
$$(-32) \times (-1) = (+32)$$
$$(-2) \div (-1) = (+22)$$
$$458 \div (-1) = -458$$

## EXPONENT AND SQUARE ROOTS

*Exponents* and *square roots* are two special indicators of arithmetic operations.

### Exponents

The use of an *exponent* is a shorthand method for indicating the number of times a number is multiplied by itself. For example, the problem $5 \times 5 \times 5 = 625$ can be written as $5^4 = 625$, which is read, "five to the fourth power equals six hundred twenty-five." The number 5 is called the *base*, 4 is the exponent, and 6255 is the fourth *power* of 5. (Sometimes the exponent is called the "power" to which a number is raised. But, in fact, the powers of a number are the results of raising that number to a given exponent. Thus, the powers of 5 are 5, 25, 125, 625, ... from $5^1, 5^2, 5^3, 5^4$....)

The exponents 2 and 3 have special names. A number raised to the exponent 2 is said to be *squared*. Thus, "five squared" is written $5^2$. The "square of 5" is 25. A number raised to the exponent 3 is said to be *cubed*. "Five cubed" is written $5^3$. The "cube of 5" is 125.

The *square root* of a given number is the number that, when squared, yields the given number. The symbol $\sqrt{\phantom{x}}$ is the square root or *radical* symbol and is placed over the number. For example, "the square root of 36," expressed as $\sqrt{36}$, is 6 because $6^2 = 6 \times 6 = 36$. Finding the square root of a number involves a very cumbersome process, but it can be done easily using a square roots table or calculator.

### CALCULATOR USE

When math problems involve negative numbers, the (+/−) key can be used to enter the sign of a number. The addition problem 2 + (−5) is entered as follows:

- press the (2) key
- press the (+) key
- press the (5) key
- press the (+/−) key
- press the (=) key to display the answer, (−3)

Several keys on scientific calculators are used to enter exponents, including the ($x^2$) and ($y^x$) keys. To multiply a number by itself, enter the number and then press the ($x^2$) key. On most calculators, the

Unit 6  NEGATIVE INTEGERS, PROPERTIES OF ZERO AND ONE, EXPONENTS AND SQUARE ROOTS  31

calculation is immediate; there is no need to press the equal (=) key. To raise a number to a power greater than 2, enter the number, press the ($y^x$) key, enter the exponent, and press the equal (=) key.

Calculators that calculate roots have keys that may be marked in several ways, including ($\sqrt{\ }$), ($\sqrt[3]{\ }$), or ($\sqrt[x]{\ }$). The ($\sqrt{\ }$) key is used to calculate a square root. On calculators that use direct algebraic logic, press the ($\sqrt{\ }$) key, enter the number, and then press the (=) sign. The square root will normally be displayed without the need to press the (=) key. The ($\sqrt[3]{\ }$) key is used to calculate a *cube root* (the number that gives the cube when raised to an exponent of 3), while the ($\sqrt[x]{\ }$) key is used to find roots other than square or cube roots.

Be sure to review your calculator instruction manual to determine how the keys are marked and the proper order in which numbers should be entered for negative numbers, exponents, and square roots.

**PRACTICAL PROBLEMS**

1. Solve the following signed addition problems.
    a. (−25) + (−48)  _____
    b. (−102) + (−1,347)  _____
    c. (−26) + (+596)  _____
    d. (+256) + (−2,331)  _____

2. Solve the following signed subtraction problems.
    a. (+8) − (+2)  _____
    b. (−29) − (−24)  _____
    c. (+375) − (−150)  _____
    d. (+450) − (+500)  _____

3. Solve the following signed multiplication problems.
    a. (+3) × (−5)  _____
    b. (−12) × (+32)  _____
    c. (−6) × (−100)  _____
    d. (−37) × (−23)  _____

4. Solve the following signed division problems.
    a. (+135) ÷ (−9)  _____
    b. (−2,500) ÷ (+50)  _____

c. (−246) ÷ (−13)

d. (+722) ÷ (−2)

5. Solve the following operations containing zero.

   a. 25 + 0

   b. (−549) − 0

   c. (−387) × 0

   d. 0 ÷ 236

6. Solve the following operations containing one.

   a. 2,567 × 1

   b. 1 × (−759)

   c. (−759) ÷ −1

   d. −826 ÷ (−1)

7. Provide a concise summary of the rules for arithmetic operations with signed numbers. _____

8. When the number 28 is squared, what is the exponent and the result?

9. Write out and solve the multiplication problem $3^3$.

10. What is "5 to the sixth power"?

11. Does $\sqrt{16} = 4$? Why or why not?

12. What does square root mean?

13. During a winter emergency, the early morning temperature is −8°F. The forecast predicts that by midday the temperature will increase by only 10°F What will the temperature be if the forecast is correct?

14. A fire department's operating budget includes a petty cash account of 2,000 dollars. If the year-end analysis indicates the account had 2,350 dollars of expenses, what is the status of the account?

15. Fire pump operators must properly calculate pressures to develop proper firestreams and to ensure the safety of personnel and equipment. When the elevation of hose lines is raised or lowered relative to the pump, a corresponding change in pressure will occur. To estimate the pressure change, fire ground hydraulic calculations typically add 5 psi for every increase in floor level and subtract 5 psi for every decrease in floor. If a hose line is advanced 4 floor levels below the apparatus (−4th floor), what is the change in pressure? _____

# Unit 7  COMBINED OPERATIONS

## OBJECTIVE

Upon completion of this unit, the student should be able to

- solve problems involving one or more arithmetic operations.

## BASIC PRINCIPLES FOR COMBINED OPERATIONS

Practical math problems often include more than one arithmetic operation within the problem. There are several considerations for solving problems with combined arithmetic operations. First, deciding which arithmetic operation to complete first, second, third, and so on, is important. The correct order of arithmetic operations must be followed to properly solve the problem. Second, arithmetic operations have three laws that can be used to help simplify and solve combined operation problems. Finally, many combined operation problems are presented as word problems. Understanding basic concepts for interpreting word problems will help ensure the problem is solved properly. Each of these considerations is discussed next, followed by several practical combined operations problems.

### Order of Operations

Often, solving mathematical problems requires several operations or steps. In such cases, a standardized approach should be followed. First, conduct operations as indicated by grouping symbols. *Grouping symbols* help indicate the order of operations within a problem. The grouping symbols are parentheses ( ), brackets [ ], and braces { }. The numbers in parentheses are operated on first, followed by those in brackets and then those in braces, if these are used. For example:

$$\{2 - [4 \times (9 + 1)]\} = \{2 - [4 \times 10]\} = \{2 - 40\} = -38$$

Second, compute exponents and roots. Third, conduct multiplication and division from left to right. Finally, conduct addition and subtraction from left to right. The acronym PEMDAS or the mnemonic phrase "**P**lease **E**xcuse **M**y **D**ear **A**unt **S**ally" may help with remembering the correct order of operations:

**P**arentheses (grouping symbols)
**E**xponents and Roots
**M**ultiplication
**D**ivision
**A**ddition
**S**ubtraction

**Example 1:** 14 × 2 + 3

**Solution:**	14 **x 2** + 3	because no **P**arentheses (grouping symbols) or **E**xponents are present, **M**ultiplication is the first operation
**28 + 3**	next order of operation is **A**ddition to complete this problem
31

**Example 2:** $4^2 \times 4 - 1$

**Solution:**	$4^2 \times 4 - 1$	start with **E**xponents
**16 x 4** – 1	next order of operations is **M**ultiplication and **D**ivision
**64 – 1**	**A**ddition and **S**ubtraction are next to complete the problem
63

**Example 3:** $(8 - 3) \times (25 \div 5)^2 + 40$

**Solution:**	**(8 – 3) × (25 ÷ 5)**$^2$ + 40	start with parentheses (note that two operations can be completed at the same time)
5 × **5**$^2$ + 40	complete square root
**5 × 25** + 40	next, multiplication before addition
**125 + 40**	finally, addition completes the problem
165

**Example 4:** $\{[(3+2) \times 5 -1] \div 2^2\} - (10 - 5)$

**Solution:**	$\{[$**(3+2)**$ \times 5 -1] \div 2^2\} - $**(10 – 5)**	parentheses are first
$\{[$**5 × 5** $-1] \div 2^2\} -5$	next work within brackets, multiplying before subtracting
$\{[$**25 – 1**$] \div 2^2\} - 5$	complete operations within brackets
$\{24 \div $**2**$^2\} - 5$	work within braces, exponents before division
$\{$**24 ÷ 4**$\} - 5$	complete operations within braces
**6 – 5**	complete the operation with subtraction
1

## ARITHMETIC LAWS

Three important arithmetic laws may assist in both understanding and solving problems. These are the *commutative, associative,* and *distributive laws.*

## Commutative Law

This law states that two numbers can be added or multiplied in any order with the same results. For example, adding 2 + 6 or 6 + 2 results in the same answer, 8. Multiplying 4 × 3 or 3 × 4 results in the same answer, 12.

## Associative Law

This law states that when three or more numbers are added, the numbers can be regrouped to achieve the same results. The same is true when three or more numbers are multiplied. Addition and multiplication examples are provided.

**Examples:** For addition: (working left and right sides separately)

$$-12 + (8 + 14) = (-12 + 8) + 14$$
$$-12 + 22 = -4 + 14$$
$$10 = 10$$

For multiplication:

$$-12 \times (8 \times 14) = (-12 \times 8) \times 14$$
$$-12 \times 112 = -96 \times 14$$
$$-1{,}344 = -1{,}344$$

## Distributive Law

This law states that when a given number is multiplied by the sum of two numbers, then the given number can by multiplied by each of the two numbers individually and then added. For example:

$$2 \times (-8 + 10) = (2 \times (-8)) + (2 \times 10)$$
$$2 \times 2 = -16 + 20$$
$$4 = 4$$

## WORD PROBLEMS

Word problems can seem confusing and, at times, almost impossible to decipher. Arithmetic word problems are essentially math problems presented in ordinary English in which all numerical values for a problem are provided. Here is a basic strategy for solving word problems:

First:   Read through the whole problem without stopping. This first step is to simply get an overall feel for the problem.

Second: Identify and label information that is relevant to the problem. In some cases, word problems contain information that is not part of the problem; focusing on only relevant information will help provide a clear picture of the problem.

Third: Convert the problem into one or more arithmetic operations.

Fourth: Solve the math problem using the correct order of operations.

Fifth: Recheck your work.

### CALCULATOR USE

Before attempting to use a calculator for solving math problems with more than one arithmetic operation, read the manual. Some calculators are not programmed to solve combined operations. Those that can solve combined operations may require different steps for entering the problem. Learning about your calculator will save time, energy, and frustration.

### PRACTICAL PROBLEMS

1. Solve the problem $-24 \times (6 + 8)$. Then rewrite the problem using the commutative law and solve the new problem.

2. Solve the problem $2 + [(18 \div 3) \times (6 + 2)]$ using the correct order of operation.

3. Solve the problem $(8 \times -4) \times 5$. Then rewrite the problem using the associative law and solve the new problem.

4. Solve the problem $\{[7 \times (300 - 100)] \div 10 + 4\}$ using the correct order of operation.

5. Solve the problem $25 \times (-14 + 2)$. Then rewrite the problem using the distributive law and solve the new problem.

6. An industrial accident causes a small leak in a tank containing 100 gallons of acetone. Before the local hazmat team can patch the tank, approximately 25 gallons of acetone are spilled. During salvage and repair operations, an additional 15 gallons are spilled.

   a. How much chemical was spilled?

   b. How much chemical remains in the tank?

38   Section 1   Basic Concepts

7. A student enrolled in a paramedic curriculum at a community college is paying the following fees:

   | Tuition: | $67 per semester hour |
   |---|---|
   | Campus access fee: | $5 per semester |
   | Technology fee: | $4 per semester hour for the first 5 hours only |
   | Lab fee: | $12 per semester |

   a. If the student is enrolled in 12 semester hours, what is the cost of tuition? _____

   b. If the student pays $125 for books, what is the total cost (including all fees) the student pays? _____

   c. If the student has saved $1,210 for school, what is the maximum number of semester hours the student will be able to take? (Include only cost of tuition in your calculation.) _____

8. Three firefighters are driving 500 miles to attend a regional conference on technical rescue. Two firefighters will be traveling together and will be staying three days and two nights. The third firefighter will be traveling alone and will be staying for two days and one night. The costs for attending the conference are as follows:

   | Conference fee: | 125 dollars per day |
   |---|---|
   | Hotel: | 97 dollars per night |
   | Meals: | 68 dollars per day |
   | Travel: | 100 dollars per vehicle |

   a. What is the total cost for the two firefighters traveling together? _____

   b. What is the total cost for the one firefighter traveling alone? _____

   c. What is the total cost for all three firefighters? _____

9. The EMS director is ordering new and improved individual CPR mask kits for all employees. One vendor is offering a dozen kits for 264 dollars, while another vendor is offering kits at 24 dollars each. If the kits from both vendors are the same quality, which vendor is offering a better price? _____

10. A paramedic is giving CPR to a victim at a rate of 2 breaths for every 15 chest compressions in a 15-second period. (Hint: There are 60 seconds in one minute.)

    a. How many breaths are given in a 1-minute period? _____

    b. How many chest compressions are given in a 1-minute period? _____

# Common Fractions

SECTION 2

## Unit 8 INTRODUCTION TO COMMON FRACTIONS

### OBJECTIVES

Upon completion of this unit, the student should be able to

- identify the different types of fractions.
- determine equivalency of fractions.
- raise and reduce the terms of a fraction.
- write integers as fractions.
- convert mixed numbers to improper fractions.
- convert improper fractions to mixed numbers.
- convert fractions in a group to equivalent fractions with common denominators.

### BASIC PRINCIPLES OF COMMON FRACTIONS

Integers are numbers that represent whole units of some thing or quantity. *Common fractions* are numbers that represent *parts* of integers. Common fractions fall between integers. The number line that follows illustrates the relationship between integers and fractions.

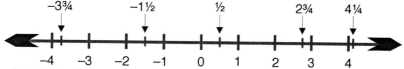

When a whole unit is divided into one or more parts, the parts are considered a fraction of the whole unit. A fraction, then, is a way of comparing a part of a whole unit and the whole. When a whole unit is divided into two equal parts, each part is called a half of a unit. When the whole unit is divided into three equal parts, each part is called a third of a unit. Whole units can be divided into any number of equal parts—fourths, fifths, sixths, and so on. The comparison of one or more parts to the whole is illustrated in the following diagram. Note that ¹⁄₁, ²⁄₂, ³⁄₃, and ⁴⁄₄ are all equivalent to one whole unit.

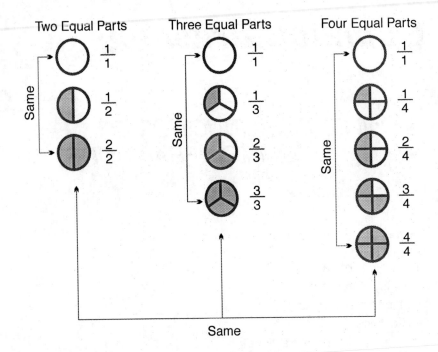

Common fractions have three components: a numerator, a denominator, and a fraction bar. The denominator (bottom number) indicates the number of equal parts into which a whole unit has been divided. The numerator (top number) indicates the number of equal parts considered in the measurement. The fraction bar is the line that separates the two numbers. The two common ways to express fractions are illustrated here. Both ways are used in this text.

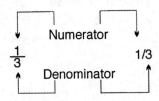

## Types of Common Fractions

A *proper fraction* has a smaller numerator than denominator. For example, ¹⁄₃₂ and ⁴⁄₅ are proper fractions. An *improper fraction* has a numerator equal to or larger than the denominator. For example, ²⁴⁄₁₀, –⁷⁄₄, and ⁸⁄₈ are considered improper fractions. *Like fractions* have the same denominator (²⁄₅ and ⁴⁄₅) while *unlike fractions* have different denominators (³⁄₈ and ⁴⁄₁₆). *Mixed numbers* consist of integers and fractions. For example 3¼, –10²⁄₅, and 2⁵⁄₁₆ are mixed numbers.

## EQUALITY OF FRACTIONS

One fractional value can be written in many different forms. Consider the following illustration. Each of the circles have been divided into equal parts, the first circle into two parts, the next circle into four parts, and so on. Notice that in each circle, the shaded area is exactly half of the circle. The fractions under each circle, although written differently, are equivalent. Fractions that are equal in value are called *equivalent fractions*.

Equivalent Fractions

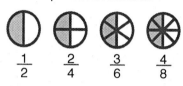

$$\frac{1}{2} \quad \frac{2}{4} \quad \frac{3}{6} \quad \frac{4}{8}$$

A simple test to determine the equivalency of two fractions is to find their *cross products*. If the cross products are equal, the fractions are equivalent. For example, the cross products for 2/4 and 4/8 are 2 × 8 = 16 and 4 × 4 = 16, as determined by the operation shown:

$$\frac{2}{4} \times \frac{4}{8} \quad \begin{array}{l} \rightarrow 4 \times 4 = 16 \\ \rightarrow 2 \times 8 = 16 \end{array}$$

Because their cross products are equal, the two fractions are equivalent.

**Example 1:** Determine if the fractions 3/7 and 12/28 are equivalent.

**Solution:**

$$\frac{3}{7} \times \frac{12}{28} \quad \begin{array}{l} \rightarrow 7 \times 12 = 84 \\ \rightarrow 3 \times 28 = 84 \end{array}$$

Because the cross products are equal, the two fractions are equivalent.

**Example 2:** Determine if the fractions 8/9 and 16/27 are equivalent.

**Solution:**

$$\frac{8}{9} \times \frac{16}{27} \quad \begin{array}{l} \rightarrow 9 \times 16 = 144 \\ \rightarrow 8 \times 27 = 216 \end{array}$$

Because the cross products are not equal, the two fractions are not equivalent.

## CHANGING COMMON FRACTION TERMS

Certain arithmetic operations require a fraction to be changed to an equivalent fraction of a higher or lower term. Changing a fraction's term means to change the form without changing the value.

Fractions can be raised to a higher term by multiplying the numerator and the denominator by the same number. The fraction 2/4 is raised to 8/16 by multiplying both the numerator and denominator by 4.

$$\frac{2}{4} \frac{\times 4}{\times 4} = \frac{8}{16}$$

To raise a fraction to a higher term with a given value in the denominator, first divide the original denominator into the new denominator. The quotient is then multiplied by the numerator and denominator of the original fraction to give the higher term fraction.

**Example:** Raise the fraction 1/3 to a higher term with a denominator of 9.

**Solution:** Since 9 ÷ 3 = 3, multiply both numerator and denominator by 3:

$$\frac{1}{3} \frac{\times 3}{\times 3} = \frac{3}{9}$$

Fractions can be reduced to a lower term by dividing the numerator and the denominator by the same number. The fraction 8/32 is reduced to 4/16 by dividing both the numerator and denominator by 2.

$$\frac{8}{32} \frac{\div 2}{\div 2} = \frac{4}{16}$$

**Example:** Reduce the fraction 2/8 to a lower term.

**Solution:**
$$\frac{2}{8} \frac{\div 2}{\div 2} = \frac{1}{4}$$

When arithmetic operations involving fractions are complete, fractions should be reduced to their lowest term. A fraction is reduced to its lowest term, simplest form, when it cannot be reduced further. The process involves repeated division by the same number or by a different number until the fraction can no longer be reduced. The greatest common factor can also be used to complete the process in one step. The greatest common factor is the largest factor that will divide both numbers without leaving a remainder.

**Example 1:** Reduce 20/32 to its lowest form using both repeated division and the greatest common factor.

**Solution:**

$$\frac{20 \div 2}{32 \div 2} = \frac{10 \div 2}{16 \div 2} = \frac{5}{8}$$

(repeated division using the same number)

$$\frac{20 \div 4}{32 \div 4} = \frac{5}{8}$$

(single step using the greatest common factor)

**Example 2:** Reduce $18/24$ to its lowest term.

**Solution:**

$$\frac{18 \div 2}{24 \div 2} = \frac{9 \div 3}{12 \div 3} = \frac{3}{4}$$

(repeated division using different numbers)

## WRITING INTEGERS AS FRACTIONS

An integer can be expressed as a fraction by placing it over a denominator of one. For example, the integer 3 can be expressed as the fraction $3/1$.

## CONVERTING MIXED NUMBERS AND IMPROPER FRACTIONS

A mixed number is converted to an improper fraction by multiplying the denominator of the fractional part by the whole number and then adding this product to the numerator. This number is then placed as the numerator over the original denominator. The mixed number $5\,3/4$ converted into an improper fraction is $23/4$, as shown by the illustration that follows:

$$5\frac{3}{4} = \frac{23}{4}$$

$(4 \times 5 + 3 = 23)$

**Example:** Convert the mixed number $4\,1/8$ to an improper fraction.

**Solution:**

$$4\frac{1}{8} = \frac{33}{8}$$

$(4 \times 8 + 1 = 33)$

An improper fraction is converted to a mixed number by dividing the denominator into the numerator. The quotient is placed as the whole number and the remainder, if any, is placed over the original denominator in the mixed number. The improper fraction 9/4 converted to a mixed number is 2¼, as shown in the following illustration:

$$\frac{9}{4} = 4\overline{)9} \quad \begin{array}{c} 2r1 \\ \phantom{4)}8 \\ \phantom{4)}\overline{\phantom{8}1} \end{array} \quad \text{or} \quad 2\frac{1}{4}$$

**Example:** Convert the improper fraction 22/5 to a mixed number.

**Solution:**

$$\frac{22}{5} = 5\overline{)22} \quad \begin{array}{c} 4r2 \\ \phantom{5)}20 \\ \phantom{5)}\overline{\phantom{2}2} \end{array} \quad \text{or} \quad 4\frac{2}{5}$$

## COMMON MULTIPLES AND DENOMINATORS

When a number is multiplied by 1, 2, 3, and so on, the product is called a *multiple* of the given number. For example, several multiples of 5 are 5, 10, 15, and 20, since

$5 \times 1 = 5$
$5 \times 2 = 10$
$5 \times 3 = 15$
$5 \times 4 = 20.$

A *common multiple* of two or more numbers is a number that is a multiple of the given numbers. For example, several common multiples of 4 and 6 are 12 and 24. These are found easily by listing the first several multiples of 4 and 6.

$4 \times 1 = 4$      $6 \times 1 = 6$
$4 \times 2 = 8$      $6 \times 2 = \mathbf{12}$
$4 \times 3 = \mathbf{12}$      $6 \times 3 = 18$
$4 \times 4 = 16$      $6 \times 4 = \mathbf{24}$
$4 \times 5 = 20$      $6 \times 5 = 30$
$4 \times 6 = \mathbf{24}$      $6 \times 6 = 36$

The *least common multiple* is the smallest common multiple of two or more numbers. In the example, it is clear that 12 is the least common multiple of 4 and 6.

When two or more fractions are involved in certain arithmetic operations, the fractions may need to be changed to equivalent fractions (fractions in which the denominators are the same). The process involves finding the *lowest common denominator* (LCD). The LCD is the least common multiple of the numbers in the denominator. If one denominator divides into the other denominator, the larger denominator is the LCD. Consider the fractions ⅓ and ¾. The least common denominator is 12, since

| | |
|---|---|
| $3 \times 1 = 3$ | $4 \times 1 = 4$ |
| $3 \times 2 = 6$ | $4 \times 2 = 8$ |
| $3 \times 3 = 9$ | $4 \times 3 = \mathbf{12}$ |
| $3 \times 4 = \mathbf{12}$ | |

The fractions ⅓ and ¾ can now be converted to equivalent fractions of 4⁄12 and 9⁄12.

$$\frac{1}{3} \quad \frac{\times 4}{\times 4} = \frac{4}{12}$$
$$\frac{3}{4} \quad \frac{\times 3}{\times 3} = \frac{9}{12}$$

Note that the cross products for ⅓ and 4⁄12 ($1 \times 12 = 12$ and $3 \times 4 = 12$) are equal, indicating the fractions are equivalent. The same is true for ¾ and 9⁄12.

The LCD for ½ and ⅛ is 8 because 2 divides 8 without leaving a remainder. This can also be seen by listing the multiples to find the least common multiple of 2 and 8.

| | |
|---|---|
| $2 \times 1 = 2$ | $1 \times 8 = \mathbf{8}$ |
| $2 \times 2 = 4$ | |
| $2 \times 3 = 6$ | |
| $2 \times 4 = \mathbf{8}$ | |

The fraction ½ is multiplied by 4⁄4 to give 4⁄8, which is an equivalent fraction of ½.

**Example:** Convert ¼ and 3⁄16 to equivalent fractions with a common denominator.

**Solution:** Since 16 is divisible by 4, 16 is the LCD.

$$\frac{1}{4} \quad \frac{\times 4}{\times 4} = \frac{8}{16}$$
$$\frac{3}{16} \quad \frac{\times 1}{\times 1} = \frac{3}{16}$$

The quickest way to find a common denominator for two fractions is to multiply together the denominators in the given fractions. This will not always give the *least* common denominator, but it will provide equivalent fractions. After writing this product in the denominators of the new fractions, proceed as if you are raising the original fractions to higher terms.

**Example:** Convert ⅕ and ⅜ to equivalent fractions with a common denominator.

**Solution:** Multiply 8 × 5 to get 40. Set up two new fractions as shown.

$$\frac{1}{5} \rightarrow \frac{\quad}{40}$$
$$\frac{3}{8} \rightarrow \frac{\quad}{40}$$

Continue by multiplying $\frac{1}{5} \times \frac{8}{8}$ and $\frac{3}{8} \times \frac{5}{5}$ to get the equivalent fractions $\frac{8}{40}$ and $\frac{15}{40}$.

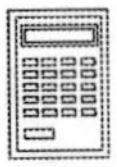

## CALCULATOR USE

If your calculator has fraction capability, be sure that you become familiar with the way your calculator handles fractions. Many scientific calculators allow fractions and mixed numbers to be entered in that format. However, the sequence of keystrokes, key symbols, and limits on the number of digits you may enter vary among calculators.

## PRACTICAL PROBLEMS

1. Provide two examples of each type of fraction.

   a. Proper fractions __________

   b. Improper fractions __________

   c. Mixed numbers __________

   d. Like fractions __________

   e. Unlike fractions __________

2. Determine if the following pairs of fractions are equivalent.

   a. $\frac{23}{48}$ and $\frac{46}{96}$ __________

   b. $\frac{1}{8}$ and $\frac{33}{264}$ __________

   c. $\frac{3}{4}$ and $\frac{42}{168}$ __________

   d. $\frac{1}{8}$ and $\frac{2}{4}$ __________

   e. $\frac{2}{5}$ and $\frac{4}{10}$ __________

For questions 3 through 5, find equivalent fractions by filling in the missing numerator or denominator.

3. $\frac{3}{4} = \frac{\quad}{16}$

4. $\frac{2}{3} = \frac{20}{\quad}$

5. $\frac{1}{5} = \frac{\quad}{15}$

For questions 6 through 10, reduce each fraction to lowest terms.

6. 5/10 _____          9. 30/48 _____

7. 45/60 _____          10. 24/36 _____

8. 8/128 _____

11. Rewrite each of the following integers as fractions.

    a. 3 _____          c. 14 _____

    b. 5 _____          d. 7 _____

For problems 12 through 16, find the lowest common denominator and convert each fraction to its equivalent with this LCD.

12. 1/4 and 3/16 _____

13. 2/3 and 4/8 _____

14. 5/8 and 11/64 _____

15. 1/8 and 3/16 _____

16. 11/32 and 4/5 _____

17. Convert 45/4 to a mixed number. _____

18. Convert 64/12 to a mixed number. _____

19. Convert 4 3/4 to an improper fraction. _____

20. Convert 8 1/2 to an improper fraction. _____

# Unit 9 ADDITION OF COMMON FRACTIONS

## OBJECTIVES

Upon completion of this unit, the student should be able to

- add common fractions.

## BASIC CONCEPTS OF COMMON FRACTION ADDITION

Addition of common fractions can only occur when the fractions to be added are like fractions. Recall that like fractions have the same denominator, which means they represent numbers that are broken into the same number of fractional parts. The addition process for fractions involves adding the fractions' numerators and keeping the common denominator. The resulting fraction should then be reduced to its lowest terms. For example, when ¼ is added to 2/4, the resulting fraction, already in its lowest terms, is ¾:

$$\frac{1}{4} + \frac{2}{4} = \frac{1+2}{4} = \frac{3}{4}$$

However, in the addition of 3/16, 2/16, and 9/16, the sum of 14/16 should be reduced to its lowest terms, which is 7/8:

$$\frac{3}{16} + \frac{2}{16} + \frac{9}{16} = \frac{3+2+9}{16} = \frac{14}{16} = \frac{7}{8}$$

**Example 1:** Add 2/4 and 3/4.

**Solution:**
$$\frac{2}{4} + \frac{3}{4} = \frac{2+3}{4} = \frac{5}{4} = 1\frac{1}{4}$$

**Example 2:** Add 3/16, 5/16, and 7/16.

**Solution:**
$$\frac{3}{16} + \frac{5}{16} + \frac{7}{16} = \frac{15}{16}$$

Before unlike fractions are added, they must first be converted to equivalent fractions with a common denominator.

**Example:** Add ¼ and 3/5

**Solution:**

$$\frac{1}{4}+\frac{3}{5}=\frac{5}{20}+\frac{12}{20}=\frac{5+12}{20}=\frac{17}{20}$$

Sometimes the sum must be reduced to its lowest terms. When adding ½ and ⅜, the equivalent fractions with a common denominator are $8/16$ and $6/16$. The sum of $14/16$ is reduced to its lowest terms, ⅞:

$$\frac{1}{2}+\frac{3}{8}=\frac{8}{16}+\frac{6}{16}=\frac{8+6}{16}=\frac{14}{16}=\frac{7}{8}$$

Note that the *least* common denominator (LCD) in this example is 8. Converting only ½ to $4/8$ and adding it to ⅜ gives ⅞. There is no need to reduce the answer when the LCD is used. The following example demonstrates this.

**Example:** Add ½ +⅓ +⅛.

**Solution:**

$$\frac{1}{2}+\frac{1}{3}+\frac{1}{8}=\frac{12}{24}+\frac{8}{24}+\frac{3}{24}=\frac{12+8+3}{24}=\frac{23}{24}$$

**ADDITION OF MIXED NUMBERS**

When adding mixed numbers, first convert the mixed numbers to improper fractions. Next, add the improper fractions as described earlier. Finally, convert the sum back to a mixed number and, if necessary, reduce it to lowest terms.

**Example 1:** Adding 6¼ to 3¼.

**Solution:** Convert mixed numbers to improper fractions:

$$6\tfrac{1}{4} = 25/4$$

$$3\tfrac{1}{4} = 13/4$$

Add the improper fractions and convert the sum to a mixed number:

$$\frac{25}{4}+\frac{13}{4}=\frac{25+13}{4}=\frac{38}{4}=9\frac{2}{4}=9\frac{1}{2}$$

**Example 2:** Add 1½ + $3\,3/34$.

50   Section 2   Common Fractions

**Solution:**

$$1\tfrac{1}{2} = \tfrac{3}{2} = \tfrac{6}{4}$$

$$3\tfrac{3}{4} = \tfrac{15}{4}$$

$$\frac{6}{4} + \frac{15}{4} = \frac{6+15}{4} = \frac{21}{4} = 5\frac{1}{4}$$

 **CALCULATOR USE**

If your calculator has fraction capability, it can be used to perform addition of fractions and mixed numbers. When entering mixed numbers on some calculators, you may need to enter a plus sign between the integer and fraction parts.

**PRACTICAL PROBLEMS**

Find the sum for each of the following. Express all answers in lowest terms.

1. $\tfrac{1}{4} + \tfrac{2}{4}$

2. $\tfrac{4}{27} + \tfrac{5}{27}$

3. $\tfrac{3}{8} + \tfrac{5}{8} + \tfrac{1}{8}$

4. $\tfrac{7}{16} + \tfrac{9}{40}$

5. $2\tfrac{1}{2}$ hours + $4\tfrac{1}{4}$ hours

6. $5\tfrac{1}{2}$ hours + $3\tfrac{1}{4}$ hours + $2\tfrac{3}{4}$ hours

7. $2\tfrac{7}{16}$ inches + 1 inch + $3\tfrac{17}{32}$ inches

8. $\tfrac{2}{3} + \tfrac{3}{4}$

9. $7\tfrac{3}{10} + 18\tfrac{4}{5} + 26\tfrac{5}{8} + 14\tfrac{3}{4}$

10. $3\tfrac{5}{8} + 20\tfrac{3}{5}$

11. During an extended incident, an emergency worker is rotated from the operations sector to the rehab (rehabilitation) sector. Upon arriving, the individual drinks ¾ cup of Gatorade. Before returning to the operations sector, the individual drinks an additional ¾ cup of Gatorade. How much Gatorade does the emergency worker drink? _____

12. Emergency responders arriving on scene quickly determine that the incident involves hazardous materials. It takes the emergency responders ¾ hour to positively identify the spilled product, 2½ hours to contain the spill, and 4⅛ hours to clean the spill. How many hours does this hazardous materials incident last? _____

13. While conducting training, firefighters practice ventilation procedures by sawing through a simulated roof with ¾-inch plywood, felt 1/16 inch thick, and a layer of shingles 3/16 inch thick. What is the total thickness of the simulated roof? _____

14. A paramedic maintains a time card for hours she works on the rescue squad. How many hours did she work for the week shown here? _____

| Day of Week | Hours Worked |
|---|---|
| Monday | 6½ |
| Tuesday | 8¼ |
| Wednesday | 0 |
| Thursday | 8½ |
| Friday | 4 5/12 |
| Saturday | 0 |
| Sunday | 4½ |
| Total | |

*52  Section 2  Common Fractions*

15. A student receives his college schedule for the first semester of study in a fire and emergency services program.

    | Course | Semester Hours |
    |---|---|
    | Communications | 2 |
    | Orientation | ¼ |
    | Study Skills | ¼ |
    | Introduction to Emergency Services | 3 |
    | First Responder | 3½ |
    | Emergency Service Safety | ¾ |
    | Physical Conditioning and Stress | 3 |

    What is the total number of semester hours? _____

16. Last week an ambulance made 5 runs. The mileage report shows the distances for each run as: 3¼, 5, 8⅛, 6, 2⁵⁄₁₆. What is the total mileage for the week? _____

17. According to the Life Safety Code, *travel distance* is the length of travel from a location in a structure to an exterior exit. What is the travel distance in feet for the illustration provided here? _____

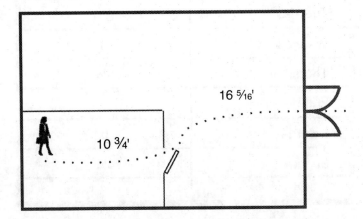

18. During a large structural fire, Engine 25 pumps 245½ gpm to an exposure line, Ladder 5 pumps 967⁵⁄₁₆ gpm through a monitor nozzle, and Engine 3 pumps 95¾ gpm through an attack line. What is the total flow for all apparatuses in gallons per minute? _____

19. From the time the alarm is received, it takes Engine 3 approximately 4¼ minutes to arrive on-scene, 8⅘ minutes to bring the fire under control, 38¾ minutes to extinguish the fire and perform salvage and overhaul, and another 20 3/16 minutes to place the apparatus back in service. What is the total duration of the incident from receipt of alarm to returning the apparatus back into service? _____

20. A shift logs the following hours of training for a month:

    Week 1          4¼
    Week 2          8 9/16
    Week 3          2⅜
    Week 4          6½

What is the month's total number of training hours for the shift? _____

# Unit 10   SUBTRACTION OF COMMON FRACTIONS

## OBJECTIVE

Upon completion of this unit, the student should be able to

- subtract common fractions.

## BASIC CONCEPTS OF COMMON FRACTION SUBTRACTION

The subtraction of common fractions is similar to the addition of common fractions. When the denominators are the same, the smaller numerator is subtracted from the larger numerator. The difference is placed as the numerator of the answer. If the fractions have different denominators, a common denominator must be found. The fractions involved are then rewritten in equivalent form with the common denominator. The difference is reduced to lowest terms, if necessary.

**Example 1:**  $12/16 - 7/16$

**Solution:**
$$\frac{12}{16} - \frac{7}{16} = \frac{12-7}{16} = \frac{5}{16}$$

**Example 2:** Subtract ¼ from ¾.

**Solution:**
$$\frac{3}{4} - \frac{1}{4} = \frac{3-1}{4} = \frac{2}{4} = \frac{1}{2}$$

**Example 3:** Subtract 3/16 from 7/8.

**Solution:**
$$\frac{7}{8} - \frac{3}{16} = \frac{14}{16} - \frac{3}{16} = \frac{14-3}{16} = \frac{11}{16}$$

### Subtracting Mixed Numbers

Before mixed numbers are subtracted, they must first be converted to improper fractions. Next, the improper fractions are subtracted as described earlier. Finally, the difference is converted back to a mixed number and, if necessary, reduced to lowest terms.

**Example:** Subtract 1⅜ from 3¼.

**Solution:** Convert mixed numbers to improper fractions with common denominators:

Unit 10 SUBTRACTION OF COMMON FRACTIONS   55

$$1\tfrac{3}{8} = \tfrac{11}{8}$$

$$3\tfrac{1}{4} = \tfrac{13}{4} = \tfrac{26}{8}$$

Subtract the improper fractions:

$$\frac{26}{8} - \frac{11}{8} = \frac{26-11}{8} = \frac{15}{8}$$

Convert the improper fraction back to a mixed number:

$$\tfrac{15}{8} = 1\tfrac{7}{8}$$

 **CALCULATOR USE**

If your calculator has fraction capability, it can be used to perform subtraction of fractions and mixed numbers. Be sure to read the operating instructions for your calculator, as different procedures are used from one calculator to another.

**PRACTICAL PROBLEMS**

1. $\tfrac{11}{16} - \tfrac{3}{16}$ _____

2. $\tfrac{15}{32} - \tfrac{11}{32}$ _____

3. ½ inch − ¼ inch _____

4. ¾ pound − ⅔ pound _____

5. $\tfrac{15}{16}$ inch − ⅜ inch _____

6. $9\tfrac{3}{4} - 6\tfrac{5}{6}$ _____

7. 1½ pounds from 4⅜ pounds _____

8. $5\tfrac{5}{16}$ yards from $25\tfrac{7}{8}$ yards _____

9. $9\tfrac{7}{16} - 3\tfrac{7}{8}$ _____

10. $\tfrac{33}{24} - \tfrac{19}{64}$ _____

56   Section 2   Common Fractions

11. A diet consultant graphs the one-month weight loss for an emergency service worker who weighs 258¾ pounds. How much does the individual weigh after 4 weeks of dieting?   _____

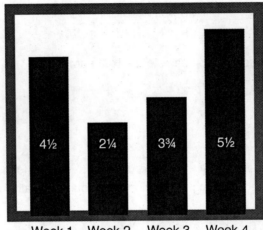

(Pounds Lost)

4½   2¼   3¾   5½

Week 1   Week 2   Week 3   Week 4

12. What is the length in inches of area A and area B on the sprinkler heads shown?

Area A: _____

Area B: _____

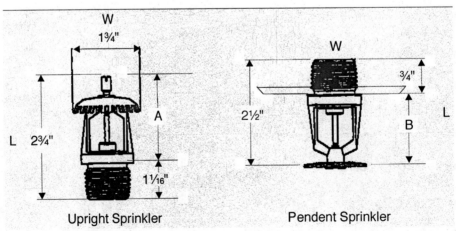

Upright Sprinkler            Pendent Sprinkler

Unit 10 SUBTRACTION OF COMMON FRACTIONS 57

13. The following is an illustration of a combination nozzle with a foam aeration attachment. The total length of both devices is 3⅞ feet. The foam adapter is 2¾ feet. How long is the nozzle (in feet)? _____

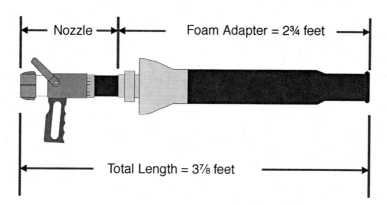

14. Total stopping distance is perception distance + reaction distance + braking distance. What is the perception/reaction distance in feet for the ambulance pictured here? _____

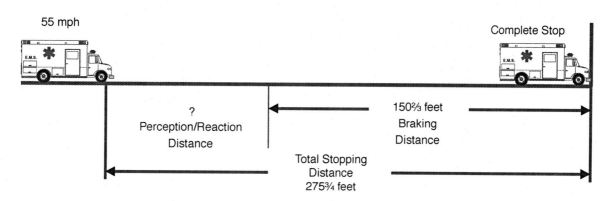

15. A training officer is designing a form to document company training activities. If 1½-inch margins are used at the top and bottom of an 11-inch sheet of paper, how much room is left for information (in inches)? _____

16. A paramedic gets overtime pay for all hours over 40 hours per week. If he works 47 5/16 hours, how many hours of overtime does he work? _____

17. A self-contained breathing apparatus (SCBA) bottle contains 45 cubic feet of air. If an emergency responder uses 24¼ cubic feet of air, how many cubic feet of air remain in the bottle? _____

18. A fire inspector spends 7½ hours conducting fire safety inspections for two structures. If the inspector spends 2¼ hours on one structure, how many hours were spent on the second structure? _____

19. During a hazardous materials incident, a hazardous material technician finds two containers with combined contents of 7⅜ gallons of hydrazine. If one container has 3¼ gallons of hydrazine, how many gallons of hydrazine does the second container hold? _____

20. Engine 4 carries 500 gallons of on-board water. To extinguish a small rubbish fire, firefighters use 200½ gallons. How much on-board water remains? _____

# Unit 11 MULTIPLICATION OF COMMON FRACTIONS

## OBJECTIVE

Upon completion of this unit, the student should be able to

- multiply common fractions.

## BASIC CONCEPTS OF COMMON FRACTION MULTIPLICATION

The multiplication of fractions is actually easier than fraction addition or subtraction. One reason for this is that common denominators are not required. The numerators of the fractions involved are multiplied, and so are the denominators, to form the product. Remember to reduce the product to its lowest terms.

**Example 1:** Multiply ¼ × ¾.

**Solution:**
$$\frac{1}{4} \times \frac{3}{4} = \frac{1 \times 3}{4 \times 4} = \frac{3}{16}$$

**Example 2:** Multiply ¾ × ⅔.

**Solution:**
$$\frac{3}{4} \times \frac{2}{3} = \frac{3 \times 2}{4 \times 3} = \frac{6}{12} = \frac{1}{2}$$

**Example 3:** Multiply ⅜ × ⅘ × ⅙.

**Solution:**
$$\frac{3}{8} \times \frac{4}{5} \times \frac{1}{6} = \frac{3 \times 4 \times 1}{8 \times 5 \times 6} = \frac{12}{240} = \frac{1}{20}$$

In some cases, the process can be simplified by *cross-cancellation*. Before multiplying, divide the numerator of one fraction and the denominator of the other fraction by a factor that is common to both. Consider the preceding example; the process is simplified as follows:

**Example:** Simplify by cross-cancellation and then multiply ⅜ × ⅘ × ⅙.

**Solution:**

$$\frac{\overset{1}{\cancel{3}}}{\underset{2}{\cancel{8}}} \times \frac{\overset{1}{\cancel{4}}}{5} \times \frac{1}{\underset{2}{\cancel{6}}} = \frac{1 \times 1 \times 1}{2 \times 5 \times 2} = \frac{1}{20}$$

(3÷3)(4÷4) above; (8÷4) (6÷3) below

## Multiplying Mixed Numbers

To multiply mixed numbers, first convert each mixed number to an improper fraction. Next, multiply the improper fractions as described earlier. Remember, simplify the fractions before multiplying, when possible, and convert the product to a mixed number.

**Example:** Multiply $2\frac{1}{4} \times 3\frac{1}{3}$.

**Solution:** $2\frac{1}{4} = \frac{9}{4}$
$3\frac{1}{3} = \frac{10}{3}$

$$\frac{\overset{3}{\cancel{9}}}{\underset{2}{\cancel{4}}} \times \frac{\overset{5}{\cancel{10}}}{\underset{1}{\cancel{3}}} = \frac{15}{2} = 7\frac{1}{2}$$

with $(9 \div 3)(10 \div 2)$ above and $(4 \div 2)(3 \div 3)$ below.

## Multiplying Fractions by Integers

When multiplying a fraction by an integer, write the integer as an improper fraction with a denominator of one.

**Example:** Multiply $3 \times \frac{7}{16}$.

**Solution:** $3 \times \frac{7}{16} = \frac{3}{1} \times \frac{7}{16} = \frac{3 \times 7}{16 \times 1} = \frac{21}{16} = 1\frac{5}{16}$

### CALCULATOR USE

If your calculator has fraction capability, fractions and mixed numbers can be multiplied. Be sure to read the operating instructions for your calculator, as different procedures are used from one calculator to another. Values may need to be entered in a specific way to ensure the order of operations is followed.

## PRACTICAL PROBLEMS

1. $\frac{1}{2} \times \frac{1}{4}$ _____
2. $\frac{1}{3} \times \frac{2}{3}$ _____
3. $\frac{3}{16} \times \frac{1}{8}$ _____
4. $\frac{7}{12} \times \frac{8}{21}$ _____
5. $9 \times \frac{1}{2}$ _____

6. $1\frac{5}{8} \times \frac{1}{3}$ _____
7. $2\frac{1}{2} \times 3\frac{1}{8}$ _____
8. $34\frac{2}{3} \times 41\frac{5}{8}$ _____
9. $12\frac{1}{2} \times \frac{2}{3}$ _____
10. $3\frac{1}{2} \times 1\frac{1}{4} \times 1\frac{3}{8}$ _____

Unit 11 MULTIPLICATION OF COMMON FRACTIONS 61

11. A paramedic knows that one small horizontal block on EKG paper represents 1/25 of a second.

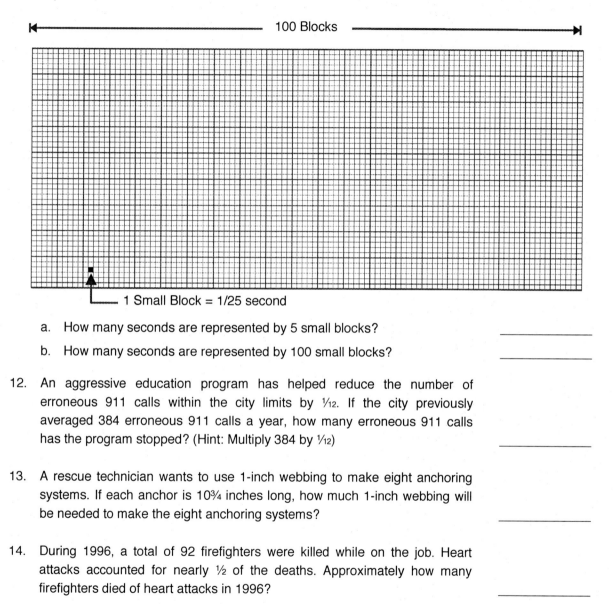

1 Small Block = 1/25 second

    a. How many seconds are represented by 5 small blocks? _____

    b. How many seconds are represented by 100 small blocks? _____

12. An aggressive education program has helped reduce the number of erroneous 911 calls within the city limits by 1/12. If the city previously averaged 384 erroneous 911 calls a year, how many erroneous 911 calls has the program stopped? (Hint: Multiply 384 by 1/12) _____

13. A rescue technician wants to use 1-inch webbing to make eight anchoring systems. If each anchor is 10¾ inches long, how much 1-inch webbing will be needed to make the eight anchoring systems? _____

14. During 1996, a total of 92 firefighters were killed while on the job. Heart attacks accounted for nearly ½ of the deaths. Approximately how many firefighters died of heart attacks in 1996? _____

15. During size-up of a structure fire, a captain uses the National Fire Academy (NFA) formula, NF (needed flow) = A/3, where A is the square foot area of the involved structure (length times width). If the structure involved in fire is 25½ feet wide by 62¾ feet long, what is the square foot area of the structure? _____

16. A county's emergency service annual budget of $80,000 is cut by ¹⁄₃₂. How much money does the budget cut represent? _____

17. A fire pump operator determines that the friction loss for 2½-inch hose flowing 240 gpm is 12½ psi per 100-foot section. If the hose line is 500 feet, what is the total friction loss for the flow in this line in psi? _____

18. A paramedic works 7½ hours a day. He spends ½ of the time in the emergency room, ¼ of the time on the ambulance, ⅛ of the time in dispatch, and ⅛ of the time in training. How many hours does he work in:

    a. the emergency room _____

    b. on the ambulance _____

    c. in dispatch and training _____

19. A county's emergency service annual budget is $450,000. If emergency operations account for ⅔ of the budget, how much is spent on emergency operations? _____

20. A hotel is 10 stories high. If each floor is 11¾ feet, how tall is the hotel? _____

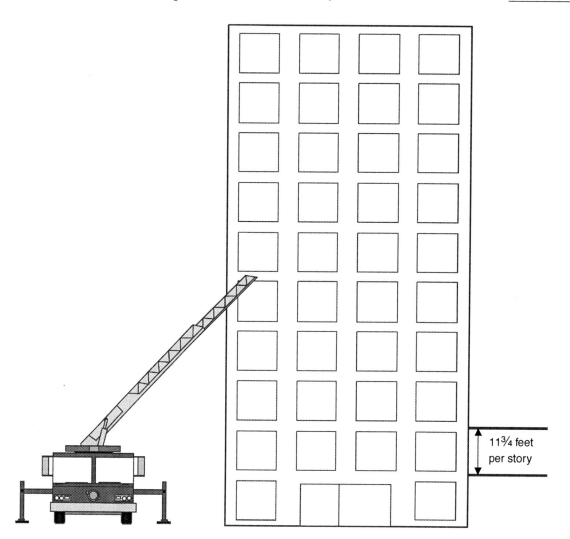

11¾ feet per story

# Unit 12  DIVISION OF COMMON FRACTIONS

## OBJECTIVE

Upon completion of this unit, the student should be able to

- divide common fractions.

## BASIC CONCEPTS OF COMMON FRACTION DIVISION

Division of common fractions involves the use of the reciprocal of a fraction. A *reciprocal* of a fraction reverses the position of the numerator and denominator. For example, the reciprocal of ¾ is 4/3. The reciprocal of 3/16 is 16/3. The reciprocal of 5 is 1/5, since 5 may be written as 5/1.

Division of common fractions is a process similar to multiplication of common fractions. The use of common denominators is not required. Before dividing fractions, the second fraction (divisor) must be converted to its reciprocal. The fractions are then multiplied. For example, to divide ¼ by ⅜, first change ⅜ to 8/3; then multiply ¼ by 8/3 and reduce the answer to lowest terms:

$$\frac{1}{4} \div \frac{3}{8} = \frac{1}{4} \times \frac{8}{3} = \frac{1 \times 8}{4 \times 3} = \frac{8}{12} = \frac{2}{3}$$

**Example:**  Divide ⅝ by ⅔.

**Solution:**
$$\frac{5}{8} \div \frac{2}{3} = \frac{5}{8} \times \frac{3}{2} = \frac{5 \times 3}{8 \times 2} = \frac{15}{16}$$

### Dividing Mixed Numbers

Before dividing mixed numbers, the mixed numbers must be changed to improper fractions. Next, the fractions are divided as described above.

**Example 1:**  Divide 2⅛ by 1½.

**Solution:**
$$2\frac{1}{8} \div 1\frac{1}{2} = \frac{17}{8} \div \frac{3}{2} = \frac{17}{8} \times \frac{2}{3} = \frac{17 \times 2}{8 \times 3} = \frac{34}{24} = \frac{17}{12} = 1\frac{5}{12}$$

**Example 2:**  Divide 3 by ⅕.

**Solution:**
$$3 \div \frac{1}{5} = \frac{3}{1} \div \frac{1}{5} = \frac{3}{1} \times \frac{5}{1} = \frac{3 \times 5}{1 \times 1} = \frac{15}{1} = 15$$

## CALCULATOR USE

If your calculator has fraction capability, it can be used to perform division operations on fractions and mixed numbers. You will need to ensure values are entered in the correct order.

## PRACTICAL PROBLEMS

1. 1/2 ÷ 1/3 _____
2. 5/8 ÷ 3/4 _____
3. 13/16 ÷ 1/8 _____
4. 9 1/2 ÷ 4 3/8 _____
5. 3/4 ÷ 60 _____
6. 3/7 ÷ 4 2/3 _____
7. 13/16 inch ÷ 1/8 inch _____
8. 5 3/8 lb ÷ 3 3/4 lb _____
9. 17 yards ÷ 5/8 yard _____
10. 25 1/4 gpm ÷ 5 1/4 gpm _____

11. An emergency medical technician (EMT) works 42 1/5 hours in a 5-day week on an ambulance. If she works the same number of hours each day, how many hours does she work per day? _____

12. A student working on a bachelor's degree in the Open Learning Fire Science program has completed 68 1/2 semester hours, or 1/2 of the required hours. How many semester hours does the student need for the degree? (Hint: Divide by fraction of total time completed.) _____

13. Two paramedics work in the emergency room of a hospital five days each week (see figure). In a one-week period, how many times longer does paramedic 2 work as compared to paramedic 1? (Hint: Divide Paramedic 2's hours by Paramedic 1's hours.)

Hours Worked per Day

| | |
|---|---|
| Paramedic 1 | 4 1/2 |
| Paramedic 2 | 6 3/4 |

14. Statistics show that about 1 out of 1,000 of all people with hepatitis die each year of fulminant hepatitis, a rapidly fatal form of the disease. If 250 people die of fulminant hepatitis each year, how many cases of hepatitis occur each year? (Hint: Write 1 out of 1,000 as the fraction 1/1,000 and use this fraction as the divisor.) _____

15. A forklift accidently runs into a tank holding 100 gallons of kerosene. The result is a 2½-gallon per minute leak. How long will it take for the tank to empty?  _____

16. One gallon of water weighs approximately 8½ pounds. If the weight of water in a tank is 11⅔ pounds, how many gallons of water are in the tank?  _____

17. The pump discharge pressure (*PDP*) for a line is 125½ psi and the nozzle pressure (*NP*) is 80¼ psi. If the total friction loss for the line is 45¼ psi, what is the average loss in pressure per 100 feet of hose if the hose line is 300 feet?  _____

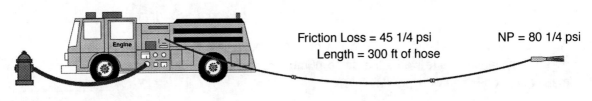

18. An EMT notes that $120 has been taken out of his paycheck for federal tax, state tax, and FICA. This is ³⁄₁₀ of his paycheck. What is his gross pay per week? (Hint: Divide $120 by ³⁄₁₀.)  _____

19. Each month during 1996 an estimated 415¾ civilian lives were lost as a result of fires. Approximately how many civilian lives were lost in a week? (Hint: Divide the number of lives lost by 4 weeks per month.)  _____

20. The weight of a gallon of water can be calculated by dividing the weight of one cubic foot of water (approximately 62½ lb/ft³) by the number of gallons in a cubic foot (approximately 7½ gal/ft³). During the division processes, the units of cubic feet cancel, leaving pounds per gallon. What is the weight of one gallon of water?  _____

# Unit 13 COMBINED OPERATIONS WITH COMMON FRACTIONS

## OBJECTIVE

Upon completion of this unit, the student should be able to

- solve combined operation problems with common fractions.

## BASIC PRINCIPLES OF COMBINED OPERATIONS WITH COMMON FRACTIONS

Follow all the rules for addition, subtraction, multiplication, and division of common fractions to solve the problems in this unit. Refer to Unit 7 for proper order of operations.

## PRACTICAL PROBLEMS

Perform the following operations.

1. $\frac{1}{4} + \frac{2}{3} - \frac{1}{8}$ _____

2. $1\frac{9}{32} - (2 \times \frac{5}{64})$ _____

3. $\frac{1}{3}[10 - (2\frac{1}{4} + 1\frac{3}{8})]$ _____

4. $(\frac{5}{6} - \frac{1}{3}) \times 8$ _____

5. $3\frac{3}{4} \times 3\frac{1}{3} \div 2\frac{1}{2}$ _____

6. $\frac{1}{4} + \frac{1}{4} \times 25$ _____

7. $\frac{3}{16} + \frac{1}{8} + \frac{7}{16} - \frac{1}{2}$ _____

8. $(15 \div \frac{1}{3}) \times 3 - 7 + 38$ _____

9. $\frac{1}{4} + \frac{3}{16} + \frac{5}{8} \div 2$ _____

10. $3\frac{1}{4}$ pounds divided by $\frac{3}{4}$ pound + 8 pounds _____

68  Section 2  Common Fractions

11. Using the chart that follows, determine the number of civilian fire deaths that occurred in 1- and 2-family dwelling occupancies during 1996. _____

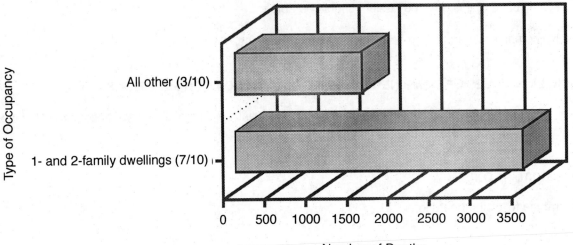

12. A fire department employs 435 personnel.

   a. If ¾ of the department personnel are firefighters, how many firefighters does the fire department employ? _____

   b. If ⅛ of the department personnel are in administration, ¼ are in training, and ¹⁄₁₆ are in supply, what is the number of personnel working in these areas combined? _____

13. An emergency medical technician (EMT) earns $8 per hour and works 8½ hours each day. A paramedic earns $12 per hour and works 7¾ hours each day. If they both work 5 days each week, what is their combined income? _____

14. The gauge on one oxygen tank shows 124³⁄₁₀ cubic feet. The second oxygen tank shows 110½ cubic feet. If 25⅕ cubic feet are used from each tank, what should each of the gauges read (that is, what is the total remaining cubic feet of oxygen in each tank)?

   Tank 1: _____

   Tank 2: _____

15. A comparison of career and volunteer fire deaths for 1996 is shown in the following chart. Determine the number of career firefighter fatalities and the number of volunteer firefighter fatalities in 1996.

    Career: _____

    Volunteer: _____

    **Comparison of On-Duty Deaths (Total Deaths = 92)**
    Career versus Volunteer (1996)
    *NFPA Journal, July/August 1997*

    Career (7/25 of total)    Volunteer (18/25 of total)

    - Fireground
    - Other On-Duty
    - Nonfire
    - Responding/Returning
    - Training

16. Three emergency vehicles travel to the scene of an accident. If one vehicle arrives in 4¼ minutes and the second in 6⅛ minutes, how long does it take the third vehicle to arrive if the sum of the travel time for each of the three vehicles is 15¹¹⁄₁₆ minutes? (Hint: Subtract the sum of the two known vehicle travel times from the total travel time of all vehicles to find the travel time of the third vehicle.) _____

17. Three fire inspectors take a total of 28¾ hours to inspect 9 buildings. If each inspector is responsible for a third of the total hours, how many hours are spent by each inspector on these buildings? _____ On average, how much time is spent on each building? _____

18. Each of three structural fires requires the following gpm flow:

    Fire #1   175⅓
    Fire #2   75¹¹⁄₁₆
    Fire #3   159½

    What is the average gpm flow for the three fires? (Hint: Add the flows and divide by the number of individual flows.) _____

19. The fire department medical director indicates in a report to the fire chief that the budget for the ambulance program needs to be increased by ⅛ over last year's budget of $25,000. The report also indicates that the new budget total is broken down as follows:

    ⅛   Operating (supplies, facility maintenance, and so on)
    ¼   Equipment (capital purchases)
    ½   Personnel (salaries/benefits)
    ⅛   Training (reimbursement for higher education)

    What is the proposed budget total and the totals for each budgeted area?

    Proposed Budget Total: _____

    Operating: _____

    Equipment: _____

    Personnel: _____

    Training: _____

Unit 13 COMBINED OPERATIONS WITH COMMON FRACTIONS  71

20. For the illustration that follows, the total gpm flowing from the pumper is 700 gpm. If one line is supplying ½ of the total gpm flow and the remaining two lines are each providing ¼ of the total gpm flow, what is the gpm flow through each line?  _____

# *Decimal Fractions*

SECTION 3

## Unit 14 INTRODUCTION TO DECIMAL FRACTIONS

**OBJECTIVES**

Upon completion of this unit, the student should be able to

- identify decimal fractions.
- convert decimal fractions to common fractions.
- round common fractions to a given place value.

**BASIC PRINCIPLES OF DECIMAL FRACTIONS**

A decimal fraction, like a common fraction, represents a part of a whole unit. Decimal fractions represent units in multiples of 10. The word *decimal* comes from a Latin word "decima," meaning tenth part. The number line that follows illustrates some relationships between common fractions and decimal fractions.

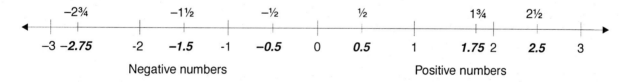

74   Section 3   Decimal Fractions

Similar to common fractions, decimal fractions have three components. For a decimal fraction, there is a whole number part, a decimal fraction part, and a decimal point. For example, in the number 4.6258, the 4 is a whole number representing four full units of something; the numbers 6258 following the decimal point represent a portion or fraction of the next whole number, which makes the number less than 5. The first number after the decimal point, 6, is in the tenths position. The following figure shows the names for the first four decimal positions.

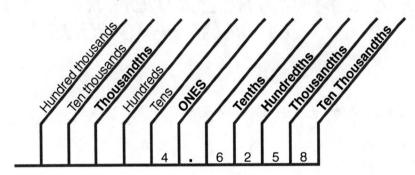

The decimal fraction 4.6 is read 4 and 6 tenths. Because the decimal fraction part represents multiples of 10, it can easily be converted to a common fraction. The decimal fraction 4.6 is converted to the common fraction 4⁶⁄₁₀, which can be reduced to 4⅗. The decimal fraction 4.62 is read 4 and 62 hundredths and is converted to the common fraction 4⁶²⁄₁₀₀, which can be reduced to 4³¹⁄₅₀. To convert the common fraction back to a decimal fraction, simply divide the numerator by the denominator. Thus, for the common fraction 4³¹⁄₅₀, divide 31 by 50 to give 0.62, which results in the decimal fraction 4.62.

**Example 1:**   Convert 23¾ to a decimal fraction.

**Solution:**   First divide:

$$\begin{array}{r} 0.75 \\ 4\overline{)3.00} \end{array}$$

The answer is 23.75

**Example 2:**   Convert 50.375 to a common fraction.

**Solution:**   Since the last digit falls in the thousandths place, write:

$$50.375 = 50\frac{375}{1000} = 50\frac{3}{8} \text{ (reduced)}$$

Zeros placed before or after the decimal fraction do not change the value of a number. For example, .6 can also be written as 0.6, 0.60, or 0.600. Often, a zero is placed before a decimal point to avoid misreading a number as an integer. For example, .6 is usually written as 0.6 to prevent misreading the number as the integer 6. Also, placing zeros to the right of a number can imply a greater degree of accuracy.

Depending on the degree of accuracy required, decimal fractions are often rounded to a specific place value. Recall from Unit 1 that to round a number, the first digit to the right of the specified place value is referred to. With decimal fractions, if this digit is less than 5, it and the remaining numbers to its right are dropped. If the digit is greater than or equal to 5, the digit in the specified place value is increased by one, and the remaining digits are dropped.

**Example 1:** Round 35.34295 to the nearest hundredths (two places).

**Solution:** 35.34 (Because 2 is less than 5, the digits to the right of 4 are dropped.)

**Example 2:** Round 28.475691 to the nearest thousandths (three places).

**Solution:** 28.476 (Because 6 is greater than 5, one is added to 5 and the digits to the right of 6 are dropped.)

Sometimes, when a common fraction is converted to a decimal fraction, it is necessary to round the answer to a desired number of places.

**Example:** Convert 5⅔ to a decimal fraction.

**Solution:** First divide:

$$\phantom{3)}\underline{0.666}$$
$$3\overline{)2.000}$$

Round to 2 places: 0.67

The answer is 5.67.

## CALCULATOR USE

Calculators are limited with regard to the number of digits that can be displayed. For example, many calculators display from eight to ten digits, although internally the calculator may store more digits. Different calculators handle rounding in different ways. Some calculators allow you to set the number of decimal places and will automatically round to the specified place value, or they will simply drop the extra digits. One technique to help avoid rounding errors is to enter the entire problem as a series of keystrokes, rather than performing partial calculations and rounding each result, then re-entering this result in other calculations. Be sure that you read the manual to understand how to display the desired number of digits.

76  Section 3  Decimal Fractions

**PRACTICAL PROBLEMS**

For problems 1 to 5, convert to equivalent forms of decimal fractions or common fractions.

1. A 1.75-inch attack line flowing 95.5 gpm _____

2. A 2.5-inch supply line flowing 225.3125 gpm _____

3. Hazmat operation lasting 4⅛ hours _____

4. Pulse rate of 65⅜ beats per minute _____

5. 225¼ ml of normal saline solution _____

For problems 6 and 7, round to the nearest tenths.

6. 5.843 _____

7. 3.457 _____

For problems 8, 9, and 10, round to nearest hundredths.

8. 245.338499 _____

9. 2,365.923999 _____

10. 0.335455 _____

# Unit 15  ADDITION OF DECIMAL FRACTIONS

## OBJECTIVE

Upon completion of this unit, the student should be able to

- add decimal fractions.

## BASIC PRINCIPLES OF ADDITION OF DECIMAL FRACTIONS

The process for adding decimal fractions is similar to that used in the addition of integers. The numbers are written in a column with the decimal points aligned. Keeping the decimal points aligned ensures the same place values are being added for each number. It may help to add zeros so all numbers have the same number of decimal places. Next, add the numbers following the same rules used for addition of integers.

**Example 1:** Add 5.33 + 23.85 + .0034.

**Solution:**

$$\begin{array}{r} \overset{1}{5}.3300 \\ 23.8500 \\ +.0034 \\ \hline 29.1834 \end{array}$$

**Example 2:** Find the total gpm flow for the following four hand lines: 55.34 gpm, 214.245 gpm, 18.959 gpm, and 203.000 gpm.

**Solution:**

$$\begin{array}{r} \overset{2\,1\,1\,1}{55}.340 \\ 214.245 \\ 18.959 \\ 203.000 \\ \hline 491.544 \text{ gpm} \end{array}$$

 *CALCULATOR USE*

A calculator can be used to add decimal fractions in a manner similar to adding whole numbers. The decimal point key (.) is pressed at the appropriate place in each number. When using a calculator for addition, it is a common practice to perform the calculations twice and compare the answers. If the answers do not match, a third calculation should be done.

## PRACTICAL PROBLEMS

1. 4.5 + 3.2 + .1

2. 20 + .25 + 23.01

3. 244.32 + 500.0335

4. 4 + 1001.22 + .33348

5. 4.25 feet + 3.33 feet + 8.23 feet

6. 250.225 gpm + 245.9393 gpm

7. .00334 + .010223 + .0000003 + .003391

8. 5,309.991 + 91.009

9. 35 + .005 + 29.342

10. 4.7 + 3.9 + 5.8 + 24.64

11. The dimensions for both a double male and double female coupling adapter are illustrated in the following diagram. What is the length of the male coupling, the female coupling, and the combined length of both couplings? (Dimensions are in inches.)

    Male coupling: _____

    Female coupling: _____

    Combined: _____

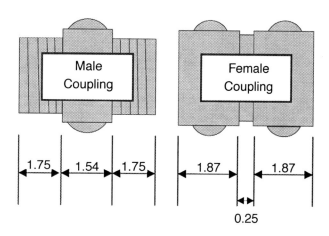

12. A firefighter analyzes the fat grams of his lunch and determines that the ham and cheese sandwich has 15.5 grams (g) of fat, the onion rings contain 12.3 g of fat, the potato chips have 7.1 g of fat, and the milkshake contains 13.8 g of fat. How many grams of fat does the meal have? _____

13. A fire protection engineer student pays $2,780 for tuition, $2,150.67 for room and board, $313.99 for books, and $24.50 for parking for one semester in college. What is the total cost for the semester? _____

14. A dietitian suggests the following breakfast for a paramedic keeping track of sodium intake:

    | | |
    |---|---|
    | bran cereal with fruit | 0.12 g of sodium |
    | 1 cup 2% milk | 0.122 g |
    | 1 muffin | 0.37 g |
    | 1 glass orange juice | 0.001 g |

    How many grams of sodium are in the meal? _____

## 80 Section 3 Decimal Fractions

15. A fire chief is working on her time management skills. She analyzes her day and finds she spends her time as follows:

    | 1.75 hrs | in the office |
    | .5 hr | traveling |
    | 3.75 hrs | in staff meetings |
    | 2 hrs | in board meetings |
    | .75 hr | surfing the net |
    | .75 hr | at a business lunch |
    | .5 hr | returning phone calls |

    How many hours does the chief spend at all these activities? _____

16. During salvage and overhaul operations, four smoke ejectors are used to remove smoke from a structure. The cubic feet per minute (cfm) rating for each ejector is: 3,700.45 cfm; 5,200.67 cfm; 9,500.14 cfm; and 10,800.97 cfm. What is the total cfm for the four ejectors? _____

17. During a wilderness search and rescue incident, six rescue technicians set out to help find a missing person. Working in pairs, they travel the following distances:

    | Team A | 3.2 miles per person |
    | Team B | 5.8 miles per person |
    | Team C | 2.9 miles per person |

    What is the combined distance traveled for each team? What is the combined distance traveled for all the rescue technicians?

    Team A: _____

    Team B: _____

    Team C: _____

    Combined: _____

18. The NFPA Journal, November/December 1997 issue, presents firefighter injury statistics for 1996. Firefighter injuries by type of duty are illustrated in the following figure. How many firefighters were injured in 1996? _____

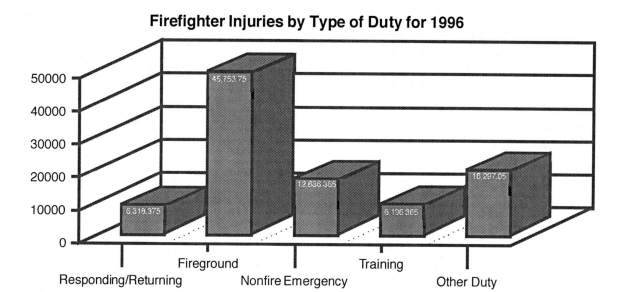

**Firefighter Injuries by Type of Duty for 1996**

- Responding/Returning: 6,318.375
- Fireground: 45,753.75
- Nonfire Emergency: 12,636.365
- Training: 6,196.365
- Other Duty: 16,297.05

19. Fire department supply personnel are ordering the equipment listed in the chart that follows. What is the total cost of the order? _____

| Item No. | Description | Unit Cost | Unit Quantity | Item Total |
|---|---|---|---|---|
| 1 | Forged Steel Shovel | $37.50 | 5 | $187.50 |
| 2 | NFPA Compliant Firefighter Gloves | $34.25 pair | 25 | $856.25 |
| 3 | Chemical Spill Pillows | $85.75 per case | 4 | $343.00 |
| 4 | Hydrant Wrench | $17.37 | 20 | $347.40 |
| 5 | Class A Foam | $13.82 per 5 gallons | 10 | $138.20 |
| | | | **Total Cost** | |

20. Five firestreams are deployed at a structure fire. The gpm flows for the five lines are:

| Line 1: | 225.45 gpm |
| Line 2: | 310.21 gpm |
| Line 3: |  95.34 gpm |
| Line 4: | 100.22 gpm |
| Line 5: | 125.34 gpm |

What is the total gpm flow? _____

# Unit 16 SUBTRACTION OF DECIMAL FRACTIONS

## OBJECTIVE

Upon completion of this unit, the student should be able to

- subtract decimal fractions.

## BASIC PRINCIPLES OF SUBTRACTION OF DECIMAL FRACTIONS

Subtraction of decimal fractions is similar to the subtraction of integers. The numbers are written in a column with the decimal points aligned. Zeros are added so all numbers have the same number of decimal places. The numbers are then subtracted following the same rules used for the subtraction of integers. Regroup as needed, being sure to place the decimal point in the answer directly below the decimal points in the numbers being subtracted.

**Example 1:** Subtract 16.23 from 42.385.

**Solution:**
```
   3 12
  42.385
 -16.230
 -------
  26.155
```

**Example 2:** Subtract 27.852 from 37.560.

**Solution:**
```
  2 16 15 5 10
  37.560
 -27.852
 -------
   9.708
```

 ## CALCULATOR USE

A calculator can be used to subtract decimal fractions in a manner similar to subtracting integers. The decimal point key (.) is pressed at the appropriate place in each number.

## PRACTICAL PROBLEMS

1. 2.7 − .05       _____

84   Section 3   Decimal Fractions

2. 78.3 – 49.538 _____

3. 0.87 – 0.28 _____

4. 18.6 – 10.89 _____

5. 123.824 – 79.55 _____

6. 9,883.456 – 298.179 _____

7. 225.34 gpm – 125.005 gpm _____

8. 3.0005 – .00342 _____

9. 45.55669 – 2.003453 _____

10. .11223 – .0021 _____

11. A pumper operating on the fireground is discharging 521.25 gpm through two lines. If one line, flowing 225.34 gpm, is shut down, how many gallons are flowing through the remaining line? _____

12. A paramedic notes the temperature of a patient as 102.9°F. After 5 minutes, the patient's temperature is 99.7°F. What is the drop in temperature? _____

13. A supply tanker contains 1,500 gallons of water. The following quantities are pumped from the tanker at different intervals: 200.5 gallons, 134.76 gallons, 500.05 gallons, and 341.21 gallons. How many gallons of water remain in the tanker? _____

14. A 1,000-gallon tanker full of kerosene is involved in an accident. If 234.65 gallons leak from the tank, how many gallons of kerosene remain in the tanker? _____

15. A toxic gas meter is used to evaluate the air at a hazmat incident. The first reading indicates 515.21 ppm (parts per million) of an unknown vapor. In the next reading, some 15 minutes later, the meter reads 500.35 ppm. What is the difference between the two readings? _____

16. An ambulance company has an annual budget of $235,215.74. If, halfway through the year, $100,231.34 has been spent, how much money remains in the budget? _____

17. A rescue technician cuts 200 feet of rope from a spool containing 1,342.63 feet of rope. How much rope remains on the spool? _____

18. At the beginning of a weight loss program, a firefighter weighs 243.64 pounds. At the end of the program, he weighs 214.03 pounds. How much weight has the firefighter lost? _____

19. At the end of the year, the odometer reading for an emergency vehicle is 2,846.4 miles. During a vehicle inspection partway through the next year, the odometer reads 3,100.9 miles. How many miles have been added? _____

# Unit 17  MULTIPLICATION OF DECIMAL FRACTIONS

## OBJECTIVE

Upon completion of this unit, the student should be able to

- multiply decimal fractions.

## BASIC PRINCIPLES OF MULTIPLICATION OF DECIMAL FRACTIONS

The process for multiplying decimal fractions is similar to the multiplication of integers. The only difference occurs after multiplying. The number of decimal places to the right of the decimal point in both the multiplier and the multiplicand are counted; this is the number of decimal places to the right of the decimal point in the answer. If extra places are needed in the answer, zeros are added on the left of the product before positioning the decimal point.

**Example 1:** Multiply 12.15 by 2.3.

**Solution:**
```
      1
   12.15     } Three decimal places total in numbers being multiplied.
   ×2.3
   ─────
   3645
   2430
   ─────
   27.945    } Place decimal point so there are three decimal places total in product.
```

**Example 2:** Multiply .04 × .8.

**Solution:**
```
    .04      } Three decimal places total.
  × .8
   ────
   .032      } Three total after adding a zero as a placeholder.
```

### CALCULATOR USE

A calculator can be used to multiply decimal fractions in a manner similar to multiplying integers. The decimal point key (.) is pressed at the appropriate place in each number. The multiplication key is normally marked with an "x."

**PRACTICAL PROBLEMS**

1. 3.5 × 1.8 _____
2. 7.27 × 31.6 _____
3. 25.78 × 9.30 _____
4. 1,254.8 × 35.375 _____
5. 24.01 × 4 _____

For problems 6 to 10, round to the nearest ten thousandths.

6. 28.561 × 5.39 _____
7. 0.123 × 0.79 _____
8. 25 × 2.03001 _____
9. 5.10223 × .334 _____
10. 3.232 × 3.22302 _____

11. If the distance between rungs on the roof ladder illustrated in the following diagram are equal, how long is the roof ladder? _____

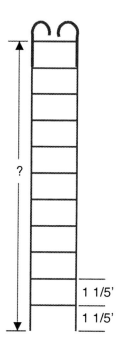

12. One millimeter (mm) distance on EKG paper equals 0.04 second. How many seconds are represented by 55 mm? _____

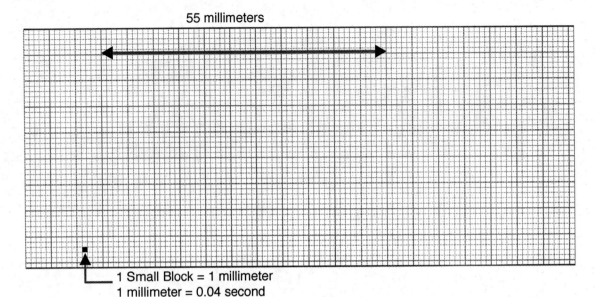

13. There are 7.48 gallons in one cubic foot of water which, in turn, weighs 62.4 lb/ft$^3$. What is the gallon capacity of a vessel containing 4,532 cubic feet of water? _____ How much does the water in the vessel weigh? _____

14. A medical supply director for an ambulance company prepares a purchase order for 50 stethoscopes. Three vendors quote prices as follows:

    Vendor A   50 stethoscopes for $712.50
    Vendor B   each package of 10 stethoscopes is $156.60
    Vendor C   each stethoscope is $15.05

    Which vendor has the lowest bid? _____

15. *Head pressure* is the vertical height of a column of liquid expressed in feet. Head pressure is calculated by multiplying 2.31 ft/psi by the pressure (expressed in psi) exerted at the base of a vessel. The formula to calculate head pressure is:

    $$h = 2.31 \text{ ft/psi} \times p \text{ (where } p = \text{pressure in psi)}$$

    If the pressure at the base of a vessel is 125 psi, what is the head pressure? (Note that the units of psi cancel, leaving ft as the unit.) _____

16. When hose lines are used at elevations above or below the fire pump, the pump operator must compensate for the resulting pressure gain or loss. To calculate this pressure gain or loss, fire pump operators multiply the height above or below the fire pump in feet by 0.5 psi. What is the pressure gain or loss for the two pumpers shown in the diagram?

    Pumper A: _____

    Pumper B: _____

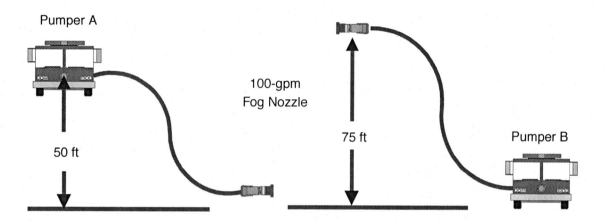

17. A fire department's annual budget will be increased by 25 percent next year. If last year's budget was $350,796.50, how much more money will be available in the new budget? _____

90   Section 3   Decimal Fractions

18. During 1996, a total of 92 firefighters lost their lives in the line of duty. Calculate the number of deaths for each cause of injury using the pie chart provided.

    Stress: _____

    Exposure: _____

    Objects: _____

    Fell: _____

    Caught/Trapped: _____

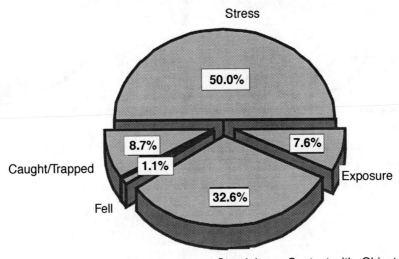

**Firefighter Deaths by Cause of Injury, 1996**

*NFPA Journal, July/August 1997*

19. The National Fire Protection Association estimates fire loss rates by region, as shown in the following chart. What is the estimated number of fires in each region for a city with a population of 250,000? (Hint: Multiply 250 by the multiplication factor.)

Northeast: _____

North Central: _____

South: _____

West: _____

| Region | Number of Fires per Thousand Population | Multiplication Factor |
|---|---|---|
| Northeast | 7.0 | .007 |
| North Central | 7.3 | .0073 |
| South | 9.0 | .009 |
| West | 5.6 | .0056 |

20. United States fire departments respond to approximately 3.75 fires every minute. How many fire responses occur in a 24-hour period? _____

# Unit 18 DIVISION OF DECIMAL FRACTIONS

## OBJECTIVE

Upon completion of this unit, the student should be able to

- divide decimal fractions.

## BASIC PRINCIPLES OF DIVISION OF DECIMAL FRACTIONS

Division of decimal fractions is like division of integers, except in the way the decimal points are treated. First, place the dividend (number to be divided) inside the division bracket and the divisor (number indicating how many times to divide the dividend) outside the bracket. Next, convert the divisor to an integer by moving the decimal point all the way to the right. Note how many places you move the decimal point, and move the decimal point of the dividend the same number of places to the right. Put a decimal point in the quotient directly over the new placement of the decimal point in the dividend. Finally, follow the same rules as for division of integers.

**Example:** Divide 13.68 by 2.4.

**Solution:**

$$2.4\overline{)13.68} \rightarrow 24.\overline{)136.8} \rightarrow 24.\overline{)136.8}^{\phantom{0}5.7}$$

$$\begin{array}{r} 120 \\ \hline 168 \\ 168 \\ \hline 0 \end{array}$$

Decimal points moved one place in divisor and dividend.

Unlike integer division, decimal fraction division does not result in a remainder. If the division does not end with a remainder of zero, zeros are added to the dividend as needed and the division is continued until the appropriate number of decimal places is achieved in the quotient. The quotient may have to be carried out to one more place than the desired precision and then rounded.

# Unit 18 DIVISION OF DECIMAL FRACTIONS

**Example:** Divide 127.05 by 8.25.

**Solution:**

$$8.25\overline{)127.05} \rightarrow 825.\overline{)12705.} \rightarrow 825\overline{)12705.0}^{\,15.4}$$

$$\begin{array}{r}\underline{825}\phantom{00}\\4455\phantom{0}\\\underline{4125}\phantom{0}\\3300\\\underline{3300}\\0\end{array}$$

Decimal points are moved two places in the divisor and the dividend. Zero is added as a placeholder to complete the division.

## CALCULATOR USE

A calculator can be used to divide decimal fractions in a manner similar to dividing integers. The decimal point key (.) is pressed at the appropriate place in each number. The division key is normally marked with (÷) or (/).

## PRACTICAL PROBLEMS

Divide the following quantities, rounding each answer to the indicated decimal place.

1. 0.96 ÷ 4 _____
2. 99.19 ÷ 0.7 _____
3. 5.7 ÷ 3.82 (two places) _____
4. 78 ÷ .007 (two places) _____
5. 30.58 ÷ 6 (one place) _____
6. 0.0057 ÷ 19 _____
7. 6.111 ÷ 0.97 (one place) _____
8. 356.76 ÷ 62 (three places) _____
9. 8.75 ÷ 1.25 _____
10. 25.05 ÷ .25 _____

11. A rescue squad purchases 15 rescue carabineers for a total of $318.75. What is the cost of each carabineer? _____

12. A paramedic works 157.5 hours in 21 days. If he works the same number of hours each day, how many hours does he work per day? _____

13. A tank containing 3,761.25 gallons will be emptied by a line that flows 250.75 gallons per minute. How long will it take to empty the tank? _____

14. A firefighter on a wellness program loses 14.96 pounds in 4 weeks. If he loses an equal number of pounds each week, how much weight does he lose each week? _____

15. Engine 2, Engine 4, and Engine 7 drop a total of 1,450 feet of supply hose. If each engine drops the same amount of hose, how much hose does each engine drop? _____

16. An emergency service dispatcher earns $8.37 per hour. One month she earns $1,255.50. Approximately how much does she make per week? _____

    How many hours per week does she work? _____

17. A 1¾-inch double jacket hose costs $107.15 per 50-foot section. If a fire department spends $2,357.30 on the hose, how many 50-foot sections does it purchase? _____

18. The National Fire Academy (NFA) formula for determining needed flow is: NF (needed flow) = A (area in square feet)/3 (constant in square feet per gpm). If the area (A) is 1,837.69 ft$^2$, what is the needed flow (NF)? (Round to two decimal places; the units of square feet cancel, leaving gpm as the unit.) _____

19. Pressure (P) is determined by dividing force (F) by area (A). If $F = 62.4$ lb/ft$^2$ and $A = 144$ in$^2$/ft$^2$, what is P? (Hint: In this example, when F is divided by A, the units ft$^2$ cancel, leaving lb/in$^2$ as the unit.) _____

20. What is the hourly rate of pay for an EMT who receives $58.50 for working 6 hours? _____

 # Unit 19 DECIMAL AND COMMON FRACTION EQUIVALENTS

## OBJECTIVES

Upon completion of this unit, the student should be able to

- convert decimal fractions to equivalent common fractions.
- convert common fractions to equivalent decimal fractions.

## BASIC PRINCIPLES OF DECIMAL AND COMMON FRACTION EQUIVALENTS

In some cases, both common fractions and decimal fractions may be included in a problem. In such cases, it is necessary to convert one or more numbers to equivalent forms so that all numbers in the problem are of the same type.

To convert a common fraction to a decimal fraction, divide the numerator by the denominator. It may be necessary to add zeros so the required number of decimal places is achieved. When this occurs, the number is usually rounded to the required decimal value.

**Example:** Convert ¾ to a decimal fraction.

**Solution:**

$$\frac{3}{4} = 4\overline{)3.00} \quad \begin{array}{r} .75 \\ \underline{28} \\ 20 \\ \underline{20} \\ 0 \end{array}$$

To convert decimal fractions to common fractions, the number to the right of the decimal point becomes the numerator with the denominator a power of 10 (10, 100, 1000, . . .). The place value of the numerator is determined by the place value of the last digit after the decimal point. For example, if the last digit to the right of the decimal is in the thousandths place, then 1,000 is placed in the denominator. The fraction is then reduced.

**Example:** Convert 2.625 to a common fraction.

**Solution:** $2.625 = 2\dfrac{625}{1000} = 2\dfrac{625 \div 125}{1000 \div 125} = 2\dfrac{5}{8}$ (reduced)

96  Section 3  Decimal Fractions

**PRACTICAL PROBLEMS**

Convert the common fractions in problems 1 through 5 to decimal fractions. Round off answers to a maximum of three places.

1. 1/8 _____

2. 50 3/16 _____

3. 43 11/16 _____

4. 250 1/4 _____

5. 4 5/32 _____

Convert the decimal fractions in problems 6 through 10 to common fractions.

6. 23.8 _____

7. 0.1875 _____

8. 51.075 _____

9. .75 _____

10. 5.25 _____

11. Express the temperature 98.6°F (Fahrenheit) as a common fraction. _____

12. Convert 250.125 gpm to a common fraction. _____

13. A building has 2,345.1875 square feet. Express this area as a common fraction. _____

14. A hose line discharges water 53.025 feet. Express this distance as a common fraction. _____

15. A smooth bore nozzle measures 1.125 inches. Express this measurement as a common fraction. _____

16. Convert the decimal fractions to common fractions for the pie chart that follows.

   Firefighter: _____

   Company Officer: _____

   Chief Officer: _____

**Firefighter Fatalities by Rank, 1996**
Total of 92 Fatalities

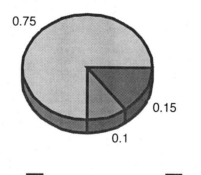

0.75

0.15

0.1

■ Firefighter    ■ Company Officer    ■ Chief Officer

# Unit 20 COMBINED OPERATIONS WITH DECIMAL FRACTIONS

**OBJECTIVE**

Upon completion of this unit, the student should be able to

- solve combined decimal fraction operations.

**BASIC PRINCIPLES OF COMBINED OPERATIONS WITH DECIMAL FRACTIONS**

Follow all the rules for addition, subtraction, multiplication, and division of decimal fractions to solve the problems in this unit.

**PRACTICAL PROBLEMS**

1. $25.47 + 5.32 - 4.3 + 3.04$ _____

2. $6.4 \div 4 + 3 \times 2.75$ _____

3. $(0.25 \times 5) + 250 \div 5$ _____

4. $3/5 \times 50 - 4.25$ _____

5. $3.005 \times 2\frac{1}{4}$ _____

6. A hazmat technician orders the equipment listed in the following chart. What is the total cost? _____

| Item Number | Quantity | Description | Unit Price | Total Cost |
|---|---|---|---|---|
| 1 | 5 | Polypropylene Nonsparking Shovel | $35.20 | |
| 2 | 10 | Chemical-Resistant Nonsparking Scoops, 82 oz | $4.40 | |
| 3 | 3 | Drum Repair Kit | $240.75 | |
| 4 | 1 | Chlorine Emergency Kit A | $1,750.25 | |
| | | | Total | |

7. A paramedic participating in a wellness program loses 3.1 pounds the first week, gains 2.45 pounds the second week, loses 2.05 pounds the third week, loses 4.3 pounds the fourth week, and gains 1.75 pounds the fifth week. If her original weight was 175¾ pounds, what is her weight at the end of the five weeks? _____

8. A normal body temperature is 98.6 degrees Fahrenheit (°F). Use the formula °C = 5/9 × (°F − 32) to calculate the temperature in Celsius degrees (°C). _____

9. A county health department charges $5.00 for a flu shot. The vaccine serum costs the health department $0.40 per shot, and a syringe costs $0.15. If a fire department sends 235 firefighters to get a flu shot, how much will it cost the fire department? _____ How much will the county health department make? _____

## Section 3 Decimal Fractions

10. Two training officers spend the following number of hours working on a new training program.

    |      | Time (hrs) | | Total |
    | Day  | Smith | Jones | |
    | --- | --- | --- | --- |
    | 1 | 4.30 | 3.01 | |
    | 2 | 6.71 | 2.80 | |
    | 3 | 7.25 | 4.20 | |
    | 4 | 8.90 | 3.70 | |
    | Total | | | |

    a. How many hours does each training officer spend on the new training program?

    Smith: _____

    Jones: _____

    b. What is the combined total number of hours spent on the project? _____

    c. What is the average number of hours spent on the project each day? _____

# Percent, Interest, Averages, and Estimates

## SECTION 4

### Unit 21 PERCENT AND PERCENTAGES

**OBJECTIVES**

Upon completion of this unit, the student should be able to

- convert values between percent, common fractions, and decimal fractions.
- use the basic percentage formula to find rate, base, or percentage.

**BASIC PRINCIPLES OF PERCENT**

The term *percent* comes from the Latin words *per centum* meaning "per hundred." The concept of percent, then, means a number of parts per hundred equal parts of a whole that is being considered. Percent can be expressed in three ways: as a percent using the percent symbol %, as a common fraction, or as a decimal fraction. For example, consider a student who correctly answers 80 questions on a 100-question exam. Eighty is the number of parts per 100 and can be listed as 80%, 0.80, or $80/100$. During certain mathematical operations, it may be necessary and/or helpful to convert to one of the three ways of expressing percent.

**From Percent to Decimal Fraction**

It is common to express a percent as a number with a percent symbol. For example, 40 percent is written as 40%. To express such a percent as a decimal fraction, divide the number by 100 or simply move the number's decimal point two places to the left and drop the percent sign. Thus 40% = .40 (note that a whole number has an unseen decimal point on its right).

## Examples

|   75%   |   25.7%   |   5%   |
|---|---|---|

```
        .75                    .257                  .05
100) 75.00             100) 25.70             100) 5.00
     700                    200                    500
     ---                    ---                    ---
     500                    570                      0
     500                    500
     ---                    ---
       0                    700
                            700
                            ---
                              0
```

OR                     OR                     OR

75% = .75.% = 0.75     25.7% = .25.7% = 0.257     5% = .05.% = 0.05

## From Decimal Fraction to Percent

To express a decimal fraction as a percent, multiply the decimal fraction by 100, or simply move the decimal point two places to the right and add a percent sign.

### Examples

|   .88   |   .5   |   .025   |
|---|---|---|

```
   100                 100                    100
 x .88               x  .5                 x .025
 ----                -----                 ------
  800                 50.0%                  500
  800                                        200
 ----                                       -----
88.00%                                      2.500%
```

OR                     OR                     OR

.88 = .88. = 88%       .5 = .50. = 50%        .025 = .025. = 2.5%

## From Percent to Common Fraction

To express percent as a common fraction requires two steps. First, change the percent to a decimal fraction by moving the decimal two places to the left and dropping the percent sign. Next, change the decimal to a fraction (described in Unit 14) and reduce to lowest terms.

**Examples**

$$60\% = .60 = \frac{60}{100} = \frac{60 \div 20}{100 \div 20} = \frac{3}{5}$$

$$45.5\% = .455 = \frac{455}{1000} = \frac{455 \div 5}{1000 \div 5} = \frac{91}{200}$$

$$2\% = .2 = \frac{2}{100} = \frac{1}{50}$$

**From Common Fraction to Percent**

To express a common fraction as percent, divide the numerator by the denominator. In the case of mixed numbers, first convert to an improper fraction. The resulting decimal is then converted into a percent by moving the decimal point two places to the right and adding the % sign.

**Examples**

$$\frac{3}{4} = 4\overline{)3.0} = .75 = 75\%$$

$$\frac{1}{8} = 8\overline{)1.00} = .125 = 12.5\%$$

$$2\frac{1}{4} = \frac{9}{4} = 4\overline{)9.0} = 2.25 = 225\%$$

Rounding should occur as directed or as appropriate for the problem at hand.

## BASIC PRINCIPLES OF PERCENTAGE

The term *percentage* is used to indicate a part of a whole. Do not confuse percent (which has the symbol % attached) with percentage. Percentage problems involve three components, shown in the following basic formula:

$P = R \times B$

where

$P$ = **P**ercentage (part of the whole)
$R$ = **R**ate (percent of the whole, expressed as decimal)
$B$ = **B**ase (whole from which a part will be described as a percentage)

The basic formula $P = R \times B$ can be algebraically rearranged to find a missing or unknown component.

## Find Percentage (P)

When rate (R) and base (B) are provided, percentage (P) can be found by simply multiplying rate times base. Remember that rate must be expressed as a decimal.

**Example 1:** What is 15% of 200?

**Solution:** $P = R \times B$
$= 15\% \times 200$
$= .15 \times 200$
$= 30$ (which means 15% of 200 is 30)

**Example 2:** Find 5% of 250 gpm.

**Solution:** $P = R \times B$
$= 5\% \times 250 \text{ gpm}$
$= .05 \times 250 \text{ gpm}$
$= 12.5 \text{ gpm}$ (which means 5% of 250 gpm = 12.5 gpm)

**Example 3:** What is ¼% of 75 psi?

**Solution:** $P = R \times B$
$= ¼\% \times 75 \text{ psi}$ (Convert ¼% to 0.25% first.)
$= 0.25\% \times 75 \text{ psi}$ (Now convert 0.25% completely to a decimal fraction by moving the decimal point and dropping % symbol.)
$= .0025 \times 750 \text{ psi}$
$= 0.1875 \text{ psi}$

## Find Rate (R)

When a problem asks for rate (R), it is asking for a *percent*. Therefore, when R is found, it has to be converted to % form. To find rate, divide the percentage by the base. The basic formula is thus changed to $R = \frac{P}{B}$. Convert the product to a percent after performing division by moving the decimal two places to the right and adding the sign %.

**Example 1:** What percent of 24 is 6?

**Solution:**
$$R = P/B$$
$$= \frac{6}{24} = 24\overline{)6.00} = 25\%$$

(long division: .25, 48, 120, 120, 0)

**Example 2:** 20 feet is what percent of 60 feet?

**Solution:** $R = P/B$

$$= \frac{20}{60} = 60\overline{)20.000}^{.333...} = 33.3\%$$ (rounded to tenths of a percent)

**Example 3:** ½ is what percent of 1?

**Solution:** $R = P/B$

$$= \frac{\frac{1}{2}}{1} = \frac{.5}{1} = 1\overline{)\,.5\,}^{.5} = 50\%$$

### Find Base (B)

To find the base, divide the percentage by the rate (that is, the percent). Remember to first change the percent to a decimal fraction. The basic formula is $B = \frac{P}{R}$.

**Example 1:** 12 is 30% of what number?

**Solution:** $B = P/R$

$$= \frac{12}{30\%} = \frac{12}{.30} = .30\overline{)12} = 30\overline{)1200.}^{40.} = 40$$

**Example 2:** 96 is 8% of what number?

**Solution:** $B = P/R$

$$= \frac{96}{8\%} = \frac{96}{.08} = .08\overline{)96} = 8\overline{)9600.}^{1200.} = 1200$$

## Example 3: 77½ is 25% of what number?

**Solution:** $B = P/R$

$$= \frac{77\frac{1}{2}}{25\%} = \frac{77.5}{.25} = .25\overline{)77.5} = 25\overline{)7750.} = 310.0$$

$$310.$$
$$25\overline{)7750.}$$
$$\underline{75}$$
$$25$$
$$\underline{25}$$
$$0$$

## PRACTICAL PROBLEMS

1. Express the following percents as decimal fractions:

   a. 25% _____    d. 120% _____

   b. 3% _____    e. ½% _____

   c. 75% _____    f. 8½% _____

2. Express the following decimal fractions as percents:

   a. 0.35 _____    d. 0.03 _____

   b. 0.4 _____    e. 0.15 _____

   c. 0.85 _____    f. 0.075 _____

3. Express the following percents as common fractions in lowest terms:

   a. 34% _____    d. 125% _____

   b. 23 percent _____    e. 75 percent _____

   c. 40% _____    f. 5% _____

4. Express the following common fractions as percents:

   a. 3/10 _____    d. 7/8 _____

   b. 5/8 _____    e. 1¼ _____

   c. 1/10 _____    f. 2/3 _____

Unit 21 PERCENT AND PERCENTAGES 107

5. Out of 250 sprinkler heads, approximately 5 percent are rejected because of defects. How many are rejected?  _____

6. The human body contains 208 bones. The fingers and toes contain a total of 56 small bones, or phalanges. What percent of the bones of the body are phalanges?  _____

7. During a three-month period, a fire department makes 208 emergency medical calls. If this represents 80% of the calls, how many total calls does the fire department respond to during the three-month period?  _____

8. The following pie chart shows emergency room admissions for a one-month period. A total of 364 patients are admitted.

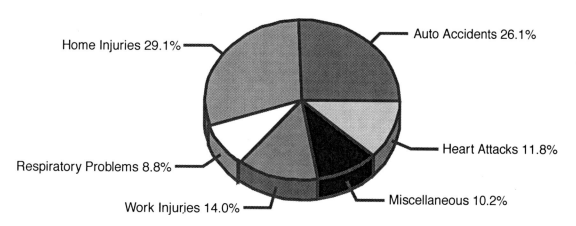

**Emergency Room Admissions**
364 Patients Total

- Home Injuries 29.1%
- Auto Accidents 26.1%
- Heart Attacks 11.8%
- Miscellaneous 10.2%
- Work Injuries 14.0%
- Respiratory Problems 8.8%

Round the answers to the following questions to the one's place.

    a. How many patients are admitted due to an injury at home or work?  _____

    b. How many people are admitted with heart attacks or respiratory problems?  _____

    c. How many more people are admitted due to automobile accidents than are admitted with heart or respiratory problems?  _____

9. An automatic sprinkler company normally includes a 1.25% cost adjustment factor in its bids. If the material is estimated to cost $12,450, what is the amount of the cost adjustment factor to be added? Round answer to hundredths place.  _____

108   Section 4   Percent, Interest, Averages, and Estimates

10. Calculate a paramedic's net weekly pay by subtracting the percentages represented by rates of deduction from the gross weekly pay. (The percentages are all taken from the original gross pay.)

|  | Gross Pay | $496.50 |
|---|---|---|
| Deduction | Rate of Deduction | Amount Deducted |
| Federal Tax | 15% |  |
| State Tax | 3.5% |  |
| City Tax | 1.5% |  |
| FICA (Social Security) | 7.65% |  |
|  | Net Pay |  |

11. A rescue technician receives a raise from $9.50 per hour to $10.25 per hour after completing a paramedic certification course. What percent (R) of his original wage is the raise? (Hint: The difference in the wages will be the percentage (P), and $9.50 is the base (B).) _____

12. Engine 21 has 1,000 feet of 4-inch supply hose. During a fire, 450 feet of the hose are used. What percent of 4-inch supply hose load is used during the fire? _____

13. How many pounds of sulfur are contained in 2,250 pounds of coal having 4% sulfur content? _____

14. A medical supply company charges 5% of the order total as a shipping/handling charge. If the order total is $775.34, what is the shipping charge? _____

15. A fire department has 355 firefighters. Over the next two years, 8¼% of the firefighters are expected to retire. If no new firefighters are hired, how many firefighters will be left in the department if the prediction occurs? _____

16. The total pressure loss in a fire hose is 14%. What is the pressure drop if the pump pressure is 100 psi? _____

17. An annual hydrostatic testing of hose finds 25% of the hose on Engine 8 has failed. If 5 sections have failed, how many sections of hose are on Engine 8? _____

# Unit 22  INTEREST AND DISCOUNTS

## OBJECTIVE

Upon completion of this unit, the student should be able to

- solve basic interest and discount problems.

## BASIC PRINCIPLES OF INTEREST

Calculating interest is one application for using percent and percentage. *Interest* is the amount of money charged by an institution to borrow money. The amount of money borrowed is called the *principal*. The *rate of interest* is the percent used to calculate the interest to be charged and is usually given for a one-year period of time. The *term* is the length of time for a loan, usually described in years or months. The *amount* (meaning amount owed) is the sum of the principal and the interest. When money is saved, interest is paid in the same manner as when money is borrowed. The basic formulas used to calculate interest and amount are:

Interest: $I = P \times R \times T$

where

- $I =$ **I**nterest
- $P =$ **P**rincipal
- $R =$ annual interest **R**ate expressed in decimal format
- $T =$ **T**ime in years or fractions of years, with common fractions converted to decimal fractions (for example, a 6-month loan = $6/12$ converted to 0.5; an 18-month loan = $18/12$ converted to 1.5)

Amount: $A = P + I$

where

- $A =$ **A**mount
- $P =$ **P**rincipal
- $I =$ **I**nterest

## Section 4  Percent, Interest, Averages, and Estimates

**Example 1:** A college student borrows $4,500 for one year at an annual rate of 4%. What is the interest and amount due at the end of the year?

**Solution:**

Interest: $I$ = $P \times R \times T$
= $4,500 \times 4\% \times 1$
= $4,500 \times .04 \times 1$
= $180

Amount: $A$ = $P + I$
= $4,500 + $180
= $4,680

**Example 2:** Find the interest and amount for a $6,200 loan at 7¼% for 3 months.

**Solution**

Interest: $I$ = $P \times R \times T$
= $6,200 \times 7¼\% \times 3/12$
= $6,200 \times 0.0725 \times .25$
= $112.375 or $112.38

Amount: $A$ = $P + I$
= $6,200 + $112.38
= $6,312.38

**Example 3:** A rescue technician invested $5,000 in a savings account for 4 years and 6 months with an annual interest rate of 7%. Find the interest earned and total amount in the savings account.

**Solution:**

Interest: $I$ = $P \times R \times T$
= $5,000 \times 7\% \times 46/12$
= $5,000 \times 0.07 \times 4.5$
= $1,575

Amount: $A$ = $P + I$
= $5,000 + $1,575
= $6,575

## BASIC PRINCIPLES OF DISCOUNTS

Calculating discounts is another common use of percent and percentage. *Discount* is the amount of money deducted from cost. *List price* is the original cost of an item. The *discount rate* is the percent by which the list price is reduced. *Net price* is the price of an item after the discount has been subtracted. The basic formulas used to calculate discount and net price are:

Discount:   $D = LP \times DR$

where

$D$ = **D**iscount
$LP$ = **L**ist **P**rice
$DR$ = **D**iscount **R**ate expressed in decimal format

Net Price:   $NP = LP - D$

where

$NP$ = **N**et **P**rice
$LP$ = **L**ist **P**rice
$D$ = **D**iscount

**Example 1:** Find the discount and net price for a purchase of $250 with a 3% discount rate.

**Solution:**

Discount:   $D = LP \times DR$
             $= \$250 \times .03$
             $= \$7.50$

Net Price:   $NP = LP - D$
             $= \$250 - \$7.50$
             $= \$242.50$

**Example 2:** The list price for a stethoscope is $32 with a 12% discount for cash payment. What is the discount and net price for a cash payment?

**Solution:**

Discount:   $D = LP \times DR$
             $= \$32 \times .12$
             $= \$3.84$

Net Price:  NP = LP − D
            = $32 − $3.84
            = $28.16

## PRACTICAL PROBLEMS

1. Calculate the interest for the following:

   a. borrowed $1,800 for 3 years with an annual interest rate of 8%  _____

   b. principal of $5,000 for a 2-year loan with an annual interest rate of 7%  _____

   c. saved $2,500 for 4 years at an annual interest rate of 3.5%  _____

2. Calculate the interest and amount for the following:

   a. borrowed $750 for 5 months at an annual rate of 12%

      Interest  _____

      Amount  _____

   b. saved $500 for 10 years at an annual rate of 8½%

      Interest  _____

      Amount  _____

3. Calculate the net price for the following:

   a. list price of $250.25 discounted 5%  _____

   b. list price of $85 discounted 20%  _____

4. An ambulance company sends out bills for $125.00, $235.25, and $267.75 on June 1 with interest charged at a rate of 8% per year. If all three bills are paid in full on December 31, with the correct amount of interest included, what is the total amount received?  _____

5. A firefighter borrows $5,000 for the purchase of a car. If the interest rate is 7%, what amount will be owed at the end of 15 months?  _____

6. A student in the paramedic degree program is buying books for the fall semester. The Campus Bookstore is selling the books for $438.52. The Pay-Less Bookstore is selling the books for $456.78 with a 5.5% discount for a cash payment. If cash is paid, which bookstore offers the better price?  _____

7. A hazardous materials technician orders $3,743.59 worth of supplies. If a 12% discount is offered for payments received within 30 days, what is the net price for the supplies? _____

8. An ambulance company borrows $14,000 for five years to upgrade computer hardware and software. If the interest rate is 7%, what is the amount of the loan? _____

9. A fire protection engineering firm is purchasing a new computer and updated software with a list price of $3,758. If the firm pays cash, a 4% discount is available. Loan options include a 1-year loan with a rate of 8.5% per year, a 2-year loan with a rate of 7.25% per year, or a 3-year loan with a rate of 4.75% per year.

    a. What is the least expensive loan? (Consider the total amount of interest paid.) _____

    b. What is the difference in cost between paying cash or taking the least expensive loan? _____

114   Section 4   Percent, Interest, Averages, and Estimates

10. An emergency rescue service makes the following purchases.

| Bill of Sale | Amounts |
| --- | --- |
| Cost of New Ambulance | $126,954.00 |
| Trade-in for Old Ambulance | $36,450.00 |
| Down Payment | $20,000.00 |
| Balance Due | |
| Interest on Balance | 9% per Year |
| Term of Loan | 2 Years |

a. What is the balance due after the trade-in and down payment are deducted?　_____

b. What is the total amount of interest paid for two years?　_____

c. If the total amount due at the end of two years is divided into monthly payments, what would each monthly payment be?　_____

# Unit 23  AVERAGES AND ESTIMATES

**OBJECTIVE**

Upon completion of this unit, the student should be able to

- calculate averages and estimates.

**BASIC PRINCIPLES OF AVERAGES**

An *average* of a set of numbers is a number that is representative of the set. One type of average is an *arithmetic mean*, found by adding all the values in the set and then dividing the sum by the number of values added. For example, the arithmetic mean of five numbers is the sum of the numbers divided by five. It is important to note that the numbers being averaged must be of the same unit of measure.

Although there are other types of averages for a set of numbers, the arithmetic mean is the calculation used most often. Thus, "average" and "arithmetic mean" will be used interchangeably in this unit.

**Example 1:** Find the arithmetic mean of 5, 10, 15, 20, and 25.

**Solution:** First add the numbers:

$$\begin{array}{r} \overset{1}{05} \\ 10 \\ 15 \\ 20 \\ +25 \\ \hline 75 \end{array}$$

Next divide the sum by the number of values (in this case, 5):

$$\begin{array}{r} 15\phantom{)} \\ 5{\overline{\smash{)}75}} \\ \underline{5}\phantom{0} \\ 25 \\ \underline{25} \\ 0 \end{array}$$

The arithmetic mean for the set of numbers is 15.

## Example 2:

Find the average gpm flow for the following hose lines:

  Line A = 75 gpm
  Line B = 90 gpm
  Line C = 125 gpm
  Line D = 250 gpm

**Solution:** Find the sum: $75 + 90 + 125 + 250 = 540$

Divide sum by number of units added: $4\overline{)540} = 135$

Average flow = 135 gpm

Averages can be used to determine an unknown quantity. Consider a student whose tests scores are 94%, 88%, and 84%. What must the student score on the fourth test to have an average of 90% for all four tests? First, multiply the desired average by the total units, and then subtract the sum of the known units to the unknown unit.

  90% (desired average) × 4 (total number of tests) = 360%
  94% + 88% + 84% (three known quantities) = 266%
  360% (desired quantity) − 266% (known quantity) = 94%

The student must earn a 94% on the fourth exam to have an overall average of 90%.

## BASIC PRINCIPLES OF ESTIMATES

*Estimates* are used for approximation purposes and are not intended to be exact. For example, a fire department records the total gallons per minute (gpm) of water used to extinguish the last seven residential fires:

| | | | |
|---|---|---|---|
| Residential Fire #1: | 3,225 gpm | Residential Fire #2: | 1,045 gpm |
| Residential Fire #3: | 550 gpm | Residential Fire #4: | 2,225 gpm |
| Residential Fire #5: | 4,000 gpm | Residential Fire #6: | 850 gpm |
| Residential Fire #7: | 1,200 gpm | | |

By adding the gpm flows together (13,095 gpm) and then dividing by the number of fires (7), it is possible to determine the average gpm flow for residential fires (1,870.7 or 1,871 gpm). Obviously, this is only an estimate, as the next residential fire will most likely require more or less than the estimated 1,870 gpm. However, when used effectively, estimating can be a powerful tool in the emergency planning process.

**PRACTICAL PROBLEMS**

1. Find the average for the following groups of numbers:

    a. 5, 14, 22, and 6 _____

    b. 25 psi, 25 psi, 23 psi, 40 psi, and 20 psi _____

    c. 25%, 15%, 50%, and 5% _____

    d. 250 gpm, 300 gpm, 500 gpm, 275 gpm, and 400 gpm _____

2. A fire science student receives the following test scores: 75%, 85%, and 95%. What is the average test score? _____

3. An emergency service student uses a modem to access the Internet through a local online provider.

    a. If the student's daily online times for one week is 30 minutes, 15 minutes, 80 minutes, 25 minutes, 45 minutes, 50 minutes, and 38 minutes, what is the average daily online time? _____

    b. If the online service costs $0.02 per minute, what is the student's average daily cost? _____

    c. What is the student's estimated monthly cost, assuming 30 days per month? _____

4. What is the average length of attack lines shown in the illustration that follows? _____

    600' of hose
    750' of hose
    350' of hose

5. The average daily sodium intake for a group of firefighters is calculated and recorded. The amounts are 2,120 milligrams (mg); 2,932 mg; 1,856 mg; 3,688 mg; 853 mg; 3,421 mg; and 1,479 mg. What is the average daily sodium intake? _____

118   Section 4   Percent, Interest, Averages, and Estimates

6. The American Heart Association estimates that 1.5 million Americans will have heart attacks in one year, and that about 35% of those stricken will die. If a comprehensive wellness program can cause a decrease in the number of heart attacks by 4% per year, how many deaths would there be from heart attacks per year at the end of the third year?   _____

7. A class of paramedics receives the test scores shown in the chart on an anatomy and physiology test. What is the average score?   _____

| Number of Students | Test Score |
| --- | --- |
| 3 | 100 |
| 3 | 97 |
| 5 | 94 |
| 6 | 90 |
| 4 | 88 |
| 1 | 82 |
| 2 | 78 |
| 3 | 75 |
| 1 | 69 |

8. A fire science student is budgeting for college. The student purchases books for four semesters as follows:

   First semester:       5 books for $432.56
   Second semester:   3 books for $296.32
   Third semester:      4 books for $337.79
   Fourth semester:    4 books for $355.21

   How much should be budgeted for books in the fifth semester if 5 books will be needed? (Hint: Find price per book for each semester and then average these amounts.)   _____

9. A hazardous materials technician is taking a chemistry course and has earned 87% and 92% on two tests, each worth 25% of the grade. If the final test is worth 50%, what grade must be earned for the student to receive a 93% as a final grade?   _____

# Measurement

SECTION 5

## Unit 24  INTRODUCTION TO MEASUREMENT

**OBJECTIVES**

Upon completion of this unit, the student should be able to

- identify the units used in the two measurement systems.
- understand the use of conversion factors for converting a measurement within and between the two systems.

**BASIC PRINCIPLES OF MEASUREMENT**

Measurement is used in every aspect of the emergency service field, from rescue to fire suppression, medical response, hazmat, and fire prevention. The two main systems of measurement used in the emergency services are the *SI (metric) system* and the *U.S. Customary system* (sometimes called the *English system*). The metric system is the most widely used measurement system in the world, while the U.S. Customary system is only used in the United States and a few smaller countries. Here are some examples of basic units for both systems:

| Measurement | U.S. | Metric |
|---|---|---|
| Length | inch (in)<br>foot (ft)<br>yard (yd)<br>mile (mi) | millimeter (mm)<br>centimeter (cm)<br>meter (m)<br>kilometer (km) |
| Volume | cup (cp)<br>pint (pt)<br>quart (qt)<br>gallon (gal) | milliliter (ml)<br>centiliter (cl)<br>liter (l)<br>kiloliter (kl) |
| Weight/Mass | ounce (oz)<br>pound (lb)<br>ton (tn) | milligram (mg)<br>centigram (cg)<br>gram (g)<br>kilogram (kg) |

Notice that the U.S. system has different units for each type of measurement: length, volume, and weight. In contrast, the metric system uses standard prefixes (milli, centi, deci, and so on) and a single basic unit for each type of measurement (meter for length, liter for volume, and gram for mass).

These measurements can be converted from one unit to another within the same system or between the two systems. The process involves developing a *conversion factor* based on equivalent measures for the units being converted. (A table of equivalent measures is provided in the appendix.) Consider the following equivalent measures and associated conversion factors.

| Equivalent Measure | Conversion Factor | |
|---|---|---|
| 1 gallon = 4 quarts | $\dfrac{4\text{ qt}}{1\text{ gal}}$ or $\dfrac{1\text{ gal}}{4\text{ qt}}$ | (Conversion within English system) |
| 1 in = 2.54 cm | $\dfrac{1\text{ in}}{2.54\text{ cm}}$ or $\dfrac{2.54\text{ cm}}{1\text{ in}}$ | (Conversion between metric and English systems) |
| 1 m = 1,000 mm | $\dfrac{1{,}000\text{ mm}}{1\text{ m}}$ or $\dfrac{1\text{ m}}{1{,}000\text{ mm}}$ | (Conversion within metric system) |

The following examples show how the conversion factors should be used. Notice that the problems are set up in a way that allows the units to cancel as needed in the multiplication process.

**Example 1:** Convert 8 gallons to quarts.

**Solution:**

In this case, choose the form of the conversion factor that allows the "gallons" to cancel. Since 8 gallons can be rewritten as $\dfrac{8\text{ gal}}{1}$, the conversion factor $\dfrac{4\text{ qt}}{1\text{ gal}}$ is used:

$$\frac{8\text{ gal}}{1} \times \frac{4\text{ qt}}{1\text{ gal}} = \frac{8\text{ gal} \times 4\text{ qt}}{1 \times 1\text{ gal}} = \frac{32\text{ qt}}{1} = 32\text{ qt}$$

**Example 2:** Convert 3.5 inches to centimeters.

**Solution:**

In this case, choose the form of the conversion factor that allows the "inches" to cancel. Since 3.5 inches can be rewritten as $\dfrac{3.5\text{ in}}{1}$, the conversion factor $\dfrac{2.54\text{ cm}}{1\text{ in}}$ should be used.

$$\frac{3.5\text{ in}}{1} \times \frac{2.54\text{ cm}}{1\text{ in}} = \frac{3.5\text{ in} \times 2.54\text{ cm}}{1 \times 1\text{ in}} = \frac{8.89\text{ cm}}{1} = 8.89\text{ cm}$$

Unit 24 INTRODUCTION TO MEASUREMENT 121

**Example 3:**

Convert 375 millimeters to meters.

**Solution:**

In this case, choose the form of the conversion factor that allows the "millimeters" to cancel. Since 375 millimeters can be rewritten as $\frac{375 \text{ mm}}{1}$, the conversion factor $\frac{1 \text{ m}}{1,000 \text{ mm}}$ should be used.

$$\frac{375 \text{ mm}}{1} \times \frac{1 \text{ m}}{1,000 \text{ mm}} = \frac{375 \cancel{\text{ mm}} \times 1 \text{ m}}{1 \times 1,000 \cancel{\text{ mm}}} = \frac{375 \text{ m}}{1,000} = 0.375 \text{ m}$$

An alternative method for converting one unit to another within the metric system is to simply move the decimal point. When changing to smaller units, shift the decimal place to the right, increasing the number of units by a power of ten; and when changing to larger units, shift the decimal place to the left, decreasing the number of units by a power of ten. The following chart can be used to indicate the direction of decimal point movement.

| Prefix | Symbol | Value |
|---|---|---|
| kilo | k | 1,000 ($10^3$) |
| hecto | h | 100 ($10^2$) |
| deka | da | 10 ($10^1$) |
| base unit |  | 1 |
| deci | d | 0.1 ($10^{-1}$) |
| centi | c | 0.01 ($10^{-2}$) |
| milli | m | 0.001 ($10^{-3}$) |

Note that the "milli" units have a value of 0.001 or $10^{-3}$. In Example 3 of the previous examples, note that the conversion from 375 mm to .375 m required the shifting of the decimal point three places to the left.

**PRACTICAL PROBLEMS**

In the following problems, find the appropriate conversion factor or convert the units as requested.

1. How many quarts are in 16 gallons? _____

2. Convert 16 quarts to gallons. _____

3. How many inches are in 30.48 centimeters? _____

4. Convert 91.44 centimeters to inches. _____

5. How many meters are in 5,525 millimeters? _____

6. Convert 5 meters to millimeters. _____

7. Convert 5 inches to millimeters. (Hint: First convert inches to centimeters, then centimeters to millimeters by moving the decimal point one place to the right.) _____

8. How many inches are in 1 meter? (Hint: First convert meter to centimeters and then centimeters to inches.) _____

9. How many meters are in 5,000 millimeters? _____

10. How many centimeters are in 36 inches? _____

# Unit 25  LENGTH MEASUREMENTS

## OBJECTIVES

Upon completion of this unit, the student should be able to

- convert length measurement within and between the two measurement systems.
- add and subtract units, including mixed units of measurement in the English system.

## LENGTH MEASUREMENT IN THE U.S. CUSTOMARY SYSTEM

Units of length in the English system are inches, feet, yards, miles, and so on. Relationships between the various units are shown in the following chart. Each of the relationships can be written as a conversion factor using the procedure described in the previous unit and shown again in the figure.

| U.S. Customary or English Length Unit | Equivalent | Conversion Factor |
|---|---|---|
| 1 ft | 12 in | $\dfrac{1 \text{ ft}}{12 \text{ in}}$ or $\dfrac{12 \text{ in}}{1 \text{ ft}}$ |
| 1 yd | 3 ft | $\dfrac{1 \text{ yd}}{3 \text{ ft}}$ or $\dfrac{3 \text{ ft}}{1 \text{ yd}}$ |
| 1 mi | 5,280 ft | $\dfrac{1 \text{ mi}}{5,280 \text{ ft}}$ or $\dfrac{5,280 \text{ ft}}{1 \text{ mi}}$ |
| 1 mi | 1,760 yd | $\dfrac{1 \text{ mi}}{1,760 \text{ yd}}$ or $\dfrac{1,760 \text{ yd}}{1 \text{ mi}}$ |

Measurement in the English system is often expressed as a series of *mixed units*, as in 6 feet 8 inches (or 6' 8"). When adding and subtracting values with mixed units, place the numbers in a column, being sure to align like units in each column. In the case of subtraction, regrouping may be necessary. When regrouping from feet to inches, for example, 1 foot may be rewritten as 12 inches and the 12 added to the inches in the original number. After completing the operation, the measurement should be written in the largest possible units.

**Example 1:** Add 4 feet 6 inches and 2 feet 8 inches.

**Solution:** First, add the values, keeping units straight.

$$\begin{array}{rr} 4\text{ ft} & 6\text{ in} \\ +\ 2\text{ ft} & 8\text{ in} \\ \hline 6\text{ ft} & 14\text{ in} \end{array}$$

The answer should then be written in the largest possible unit. Note that 14 inches is the same as 1 foot 2 inches. The sum of 6 feet 14 inches is converted to the larger unit of 7 feet 2 inches as follows:

6 ft 14 in = 6 ft + 1 ft 2 in = 7 ft 2 in

**Example 2:** Subtract 3 feet 9 inches from 6 feet 6 inches.

**Solution:** First, align the values to be subtracted.

$$\begin{array}{rr} 6\text{ ft} & 6\text{ in} \\ -\ 3\text{ ft} & 9\text{ in} \end{array}$$

Note that the inches column requires borrowing. In this case, 1 foot or 12 inches is borrowed from the feet column and added to the inches column, with 6 feet changing to 5 feet:

$$\begin{array}{rr} 5\text{ ft} & 18\text{ in} \\ -\ 3\text{ ft} & 9\text{ in} \\ \hline 2\text{ ft} & 9\text{ in} \end{array}$$

Be sure to add like units, such as feet to feet and inches to inches.

When converting from one unit to another in the English system, be sure to choose the conversion factor that cancels units as needed.

**Example 1:** How many inches are in 3 feet?

**Solution:** $\dfrac{3\text{ ft}}{1} \times \dfrac{12\text{ in}}{1\text{ ft}} = \dfrac{3\text{ ft} \times 12\text{ in}}{1 \times 1\text{ ft}} = \dfrac{36\text{ in}}{1} = 36\text{ in}$

**Example 2:** Convert 21,120 feet to miles.

**Solution:** $\dfrac{21{,}120\text{ ft}}{1} \times \dfrac{1\text{ mi}}{5{,}280\text{ ft}} = \dfrac{21{,}120\text{ mi}}{5{,}280} = 4\text{ mi}$

## LENGTH MEASUREMENT IN THE SI (METRIC) SYSTEM

The standard unit for length measurement in the SI system is the *meter*. Changing from one unit to another within the metric system can be accomplished by using a conversion factor or by moving the decimal point.

| Metric Linear Unit | Symbol | Value in Meters |
| --- | --- | --- |
| kilometer | km | 1,000 ($10^3$) |
| hectometer | hm | 100 ($10^2$) |
| dekameter | dam | 10 ($10^1$) |
| meter | m | 1 |
| decimeter | dm | 0.1 ($10^{-1}$) |
| centimeter | cm | 0.01 ($10^{-2}$) |
| millimeter | mm | 0.001 ($10^{-3}$) |

**Example 1:** Convert 450 centimeters to meters.

**Solution:**
$$\frac{450 \text{ cm}}{1} \times \frac{1 \text{ m}}{100 \text{ cm}} = \frac{450 \text{ m}}{100} = 4.5 \text{ m}$$

Alternately, because the "centi" has a value of $10^{-2}$ or 0.01, the decimal point can be moved two places to the left on 450 cm to give 4.50 m or 4.5 m.

**Example 2:** How many meters are in 30 kilometers?

**Solution:**
$$\frac{30 \text{ km}}{1} \times \frac{1,000 \text{ m}}{1 \text{ km}} = 30,000 \text{ m}$$

Alternately, because "kilo" has a value of $10^3$ or 1,000, simply move the decimal point three places to the right on 30 km to give 30,000 m.

## CONVERSION OF LENGTH MEASUREMENTS BETWEEN SYSTEMS

Conversion factors are also used to convert units between the two systems. The following relationships of units between the two systems can be used to develop conversion factors. Keep in mind that conversion factors should be developed to cancel units as needed.

| U.S. Customary Unit | SI (Metric) Equivalent |
|---|---|
| 1 in | 2.54 cm |
| 1 ft | 0.3048 m |
| 1 yd | 0.9144 m |
| 1 mi | 1.609 km |

**Example 1:** A friend is running in a 5-kilometer race. How many miles will the friend run?

Solution: $\dfrac{5 \text{ km}}{1} \times \dfrac{1 \text{ mi}}{1.609 \text{ km}} = \dfrac{5 \text{ mi}}{1.609} = 3.1 \text{ mi}$

**Example 2:** How many meters are in 50 yards?

Solution: $\dfrac{50 \text{ yd}}{1} \times \dfrac{0.9144 \text{ m}}{1 \text{ yd}} = \dfrac{45.72 \text{ m}}{1} = 45.72 \text{ m}$

## PRACTICAL PROBLEMS

For problems 1 through 9, use the proper conversion factor to obtain the requested unit.

1. Convert 2.5 meters to centimeters. _____

2. How many meters are in 3,805 millimeters? _____

3. Convert 42.5 centimeters to millimeters. _____

4. Convert 5 feet to inches. _____

5. How many yards are in 5 miles? _____

6. Convert 500 yards to feet. _____

7. Convert 15 kilometers to miles. _____

8. How many inches are in 25 centimeters? _____

9. How many meters are in 100 feet? _____

10. During a search and rescue operation, a rescue worker walks 7½ miles. How many kilometers has the rescue worker traveled? _____

11. Most fire hose is 50 feet in length. Convert that length of hose to the following units:
    a. inches _____
    b. meters _____
    c. yards _____
    d. centimeters _____

12. A heart attack patient starts an exercise program. The first day he walks 0.5 kilometers. Each day he increases his distance by 50 meters. At the end of one week (7 days), how far is he walking? _____

128  Section 5  Measurement

13. For each hose line in the illustration that follows, convert the length of hose to meters and the diameter of hose to centimeters.

   Line A— Length          _____
           Diameter        _____
   Line B— Length          _____
           Diameter        _____
   Line C— Length          _____
           Diameter        _____
   Line D— Length          _____
           Diameter        _____
   Line E— Length          _____
           Diameter        _____

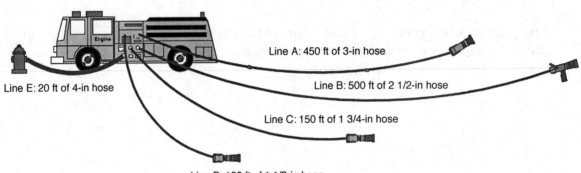

Line E: 20 ft of 4-in hose
Line A: 450 ft of 3-in hose
Line B: 500 ft of 2 1/2-in hose
Line C: 150 ft of 1 3/4-in hose
Line D: 100 ft of 1 1/2-in hose

14. A firefighter is running in a 10-kilometer run. How many miles is the run?  _____

# Unit 26  AREA AND PRESSURE MEASUREMENTS

## OBJECTIVE

Upon completion of this unit, the student should be able to

- calculate area and pressure measurement.

## BASIC PRINCIPLES OF AREA MEASUREMENT

*Area* is a measurement of the size of a *surface*. It is a two-dimensional measurement of width and length, but with no thickness or depth. Area measurement is most often expressed in square units such as square inches (in$^2$), square meters (m$^2$), or square feet (ft$^2$). Formulas for the areas of various shapes are given in the following chart. In some instances, it may be necessary to divide a space into smaller pieces, calculate individual areas, and add these smaller areas to find the total area. The lengths of the sides of an object must be in the same unit before they are multiplied.

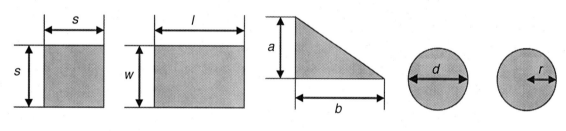

Square          Rectangle          Triangle          Circle

Square:    $A = s^2$          $s$ = length of side

Rectangle: $A = l \times w$     $l$ = length
                              $w$ = width

Triangle:  $A = \dfrac{ab}{2}$   $a$ = altitude (height)
                              $b$ = base

Circle:    $A = \pi r^2$        $\pi$ = 3.1416
                              $r$ = radius      or

           $A = \dfrac{\pi d^2}{4}$   $\pi$ = 3.1416
                              $d$ = diameter

**Example:** Calculate the area for each of the following.

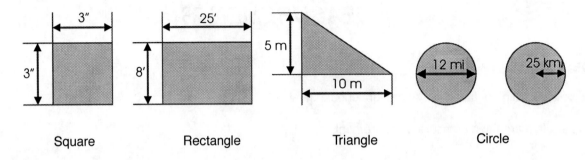

Square   Rectangle   Triangle   Circle

**Solution:**

Square: $A = s^2$ or $s \times s$
$= 3 \text{ in} \times 3 \text{ in}$
$= 9 \text{ in}^2$

Rectangle: $A = l \times w$
$= 25 \text{ ft} \times 8 \text{ ft}$
$= 200 \text{ ft}^2$

Triangle: $A = \dfrac{ab}{2}$

$= \dfrac{5 \text{ m} \times 10 \text{ m}}{2} = \dfrac{50 \text{ m}^2}{2} = 25 \text{ m}^2$

Circle: (diameter)

$A = \dfrac{\pi d^2}{4}$

$= \dfrac{3.1416 \times (12 \text{ mi} \times 12 \text{ mi})}{4} = \dfrac{3.1416 \times 144 \text{ mi}^2}{4} = \dfrac{452.39 \text{ mi}^2}{4} = 113.0976 \text{ mi}^2 \text{ or } 113.1 \text{ mi}^2$

(using formula with radius)

$A = \pi r^2$
$= 3.1416 \times (25 \text{ km} \times 25 \text{ km})$
$= 3.1416 \times 625 \text{ km}^2$
$= 1{,}963.5 \text{ km}^2$

When converting one *square unit* to another square unit, use the same conversion factor, but use it twice. For example, to convert inches to feet, use $(\dfrac{1 \text{ ft}}{12 \text{ in}} \times \dfrac{1 \text{ ft}}{12 \text{ in}})$ or $\dfrac{1 \text{ ft}^2}{144 \text{ in}^2}$.

**Example:** Calculate the area of a square 18 inches on each side in square feet.

**Solution:** First calculate area in square inches:

$$A = s^2$$
$$= 18 \text{ in} \times 18 \text{ in} = 324 \text{ in}^2$$

Next, convert from square inches to square feet:

$$\frac{324 \text{ in}^2}{1} \times \frac{1 \text{ ft}}{12 \text{ in}} \times \frac{1 \text{ ft}}{12 \text{ in}} = \frac{324 \text{ ft}^2}{144} = 2.25 \text{ ft}^2 \text{ or}$$

$$\frac{324 \text{ in}^2}{1} \times \frac{1 \text{ ft}^2}{144 \text{ in}^2} = \frac{324 \text{ ft}^2}{144} = 2.25 \text{ ft}^2$$

## BASIC PRINCIPLES OF PRESSURE MEASUREMENT

*Pressure* is force applied over an area. It is expressed in units of force per unit area. Common units of pressure include pounds per square inch (psi) or pounds per square foot in the U.S. Customary system or Newtons per square meter (pascals) in the SI system. The basic formula for calculating pressure is:

$$P = F/A, \text{ where } F = \text{force (or weight) and } A = \text{surface area}$$

**Example:** Find the pressure exerted by a 50-pound force over an area of 5 square feet.

**Solution:** $P = F/A$

$$P = \frac{50 \text{ lb}}{5 \text{ in}^2} = 10 \text{ lb per in}^2 \text{ or } 10 \text{ psi}$$

## PRACTICAL PROBLEMS

1. Find the area of a square measuring 4 meters on each side.

2. Find the area of a rectangle 4 feet 7 inches long by 3 feet 9 inches wide.

3. Find the area of a circle whose diameter is 8 centimeters.

4. Find the area of a triangle with a base length of 60 feet and a height of 25 feet.

5. Find the area of a circle with a radius of 5½ inches.

6. Find the pressure exerted by a 10-pound force over an area of 5 square inches. _____

7. What is the area of a structure measuring 75 feet long by 45 feet wide? _____

8. Find the area of a circle whose diameter is 2½ inches. _____

9. What is the pressure exerted by a 150-pound force over a circle with a 2-inch radius. (Hint: First find the area of the circle.) _____

10. For each value shown in the chart below, find the missing equivalent value as requested. Round your answer to the nearest one's place. _____

| Convert From | Convert To | Result |
|---|---|---|
| 2 ft² | in² | |
| 125 in² | cm² | |
| 9 m² | yd² | |
| 180 ft² | yd² | |

11. Find the pressure in pounds per square foot for a 6,000-pound force applied over an area 40 feet wide by 60 feet long. _____

12. For problem 11, find the pressure in pounds per square inch. _____

13. What is the square footage of a classroom measuring 38.5 feet by 2.8 feet? _____

14. The total surface area of a firefighter's boot soles is 70 square inches. The firefighter, in full protective clothing, weighs 280 pounds. If the weight is distributed equally on the soles of the boots, the pressure exerted by the firefighter's weight on the ground is how many pounds per square inch? _____
Convert the pressure to pounds per square foot. _____

15. How many square yards are in a salvage cover measuring 14 feet by 18 feet? _____

# Unit 27 SOLID AND FLUID VOLUME MEASUREMENTS

## OBJECTIVE

Upon completion of this unit, the student should be able to

- calculate solid and fluid volume measurement within and between the two measurement systems.

## BASIC PRINCIPLES OF VOLUME MEASUREMENT FOR SOLIDS

*Volume* is the amount of space occupied by an object. It is a three-dimensional measurement. When volume is measured in linear measurement, called *cubic measure*, the units are typically cubic inches ($in^3$), cubic feet ($ft^3$), and cubic meters ($m^3$). The two basic volume formulas are:

Rectangle: $V = LWH$

Cylinder: $V = Ah$ ($A = \pi r^2$) or $V = \pi r^2 h$

**Example 1:** Find the volume for the rectangle shown in the diagram.

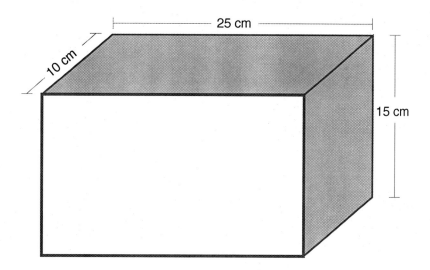

**Solution:**  $V = LWH$
= 10 cm x 25 cm x 15 cm
= 3,750 $cm^3$

**Example 2:** Find the volume of the cylinder shown in the diagram.

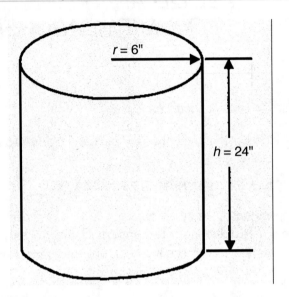

**Solution:**  $V = Ah$
$= (\pi r^2) \times h$
$= (3.1416 \times 6in^2) \times 24\ in$
$= (3.1416 \times 36\ in^2) \times 24\ in$
$= 113.0976\ in^2 \times 24\ in$
$= 2,714.34\ in^3$ or $2,714\ in^3$

When converting from one volume unit to another, use the linear conversion factor three times. For example, to convert inches to feet, use $\frac{1\ ft}{12\ in}$ three times, or $(\frac{1\ ft}{12\ in} \times \frac{1\ ft}{12\ in} \times \frac{1\ ft}{12\ in})$ or $\frac{1\ ft^3}{1,728\ in^3}$.

**Example:** Convert 300 in³ to ft³.

**Solution:**
$$\frac{300\ \cancel{in^3}}{1} \times \frac{1\ ft}{12\ in} \times \frac{1\ ft}{12\ in} \times \frac{1\ ft}{12\ in} = \frac{300\ \cancel{in^3} ft^3}{1,728\ \cancel{in^3}} = \frac{300\ ft^3}{1,728} = 0.174\ ft^3$$

or

$$\frac{300\ \cancel{in^3}}{1} \times \frac{1\ ft^3}{1,728\ \cancel{in^3}} = \frac{300\ ft^3}{1,728} = 0.1736\ ft^3$$

## BASIC PRINCIPLES OF VOLUME MEASUREMENT FOR FLUIDS

In some cases, emergency service mathematics problems involve fluid volume. In the English system, the units for liquid measurement are teaspoon, tablespoon, ounce, pint, quart, and gallon. In the metric system, the basic unit for liquid is the liter. The following chart provides standard volume measurements and equivalent measurements used in the English and metric systems. Remember, when converting from one unit to another, be sure to select the conversion factor that produces the desired units after multiplication.

| Basic Unit | Equivalents |
|---|---|
| 1 teaspoon (tsp) | 5 milliliters (ml) |
| 1 tablespoon (tsp) | 3 tsp, 15 ml |
| 1 ounce (oz) | 30 ml, 29.574 cubic centimeters ($cm^3$ or cc) |
| 1 cup (cp) | 8 oz, 240 ml |
| 1 pint (pt) | 2 cp, 480 ml |
| 1 quart (qt) | 2 pt, 0.9464 liter (l) |
| 1 gallon (gal) | 4 qt, 3.785411 liters (l), 0.13368 $ft^3$, 231$in^3$) |
| 1 ml | 1 cubic centimeter ($cm^3$ or cc) |
| 1 liter | 1,000 ml, 1,000 cc, 0.264172 gal, 1.056688 qt |
| 1 $ft^3$ | 7.48 gal |

An important equivalent measure in the metric system is 1 milliliter = 1 cubic centimeter, or 1 ml = 1 cc or $cm^3$.

**Example:** How many gallons are in a tank measuring 5 feet long, 2 feet wide, and 1 foot high?

**Solution:** You must first calculate the volume in cubic measurement.

Cubic Measurement:

$$V = LWH$$
$$= 5 \text{ ft} \times 2 \text{ ft} \times 1 \text{ ft}$$
$$= 10 \text{ ft}^3$$

Then select the appropriate conversion factor from the chart to change the cubic measurement to fluid volume.

Fluid volume measurement:

$$1 \text{ ft}^3 = 7.48 \text{ gal. or } 7.48 \text{ gal/ft}^3$$

$$\frac{10 \text{ ft}^3}{1} \times \frac{7.48 \text{ gal}}{\text{ft}^3} = 74.8 \text{ gal}$$

Alternately, use 1 gal = 0.134 $ft^3$, so $\frac{10 \text{ ft}^3}{1} \times \frac{1 \text{ gal}}{0.134 \text{ ft}^3} = 74.8 \text{ gal}$

## PRACTICAL PROBLEMS

1. What is the volume of a cube with sides 5 meters long?

2. What is the volume of a rectangle measuring 5 inches wide by 6 inches long by 2 inches high?

3. Find the volume of a rectangular solid 12.5 centimeters long, 8.25 centimeters wide, and 4.5 centimeters high. Round to nearest tenth.

4. Find the volume of a rectangular solid measuring 5 feet wide, 5 feet long, and 6 inches high. Express the volume in both cubic inches and cubic feet.

5. What volume can a 50-foot long fire hose with a 2½-inch diameter hold? (Hint: Treat the fire hose as a cylinder with the length used for height.) Express the volume in cubic feet and gallons.

6. Engine 6 carries 500 gallons of water and 100 gallons of foam. What is the equivalent amount of liters for both the water and foam?

7. How many cubic inches are in a cubic foot?

8. What is the volume, in both cubic feet and cubic meters, of a cylindrical tank measuring 6 feet in diameter by 10 feet high?

9. How many gallons of water can a cylindrical tank hold that measures 5 meters in diameter and 15 meters in height?

10. Convert the following to liters:

    a. 596 ml

    b. 5 qt

    c. 10 pt

    d. 10 cc

11. During a 24-hour period, a patient receives 2.5 liters (liters) of intravenous (IV) solution and drinks 2,440 cubic centimeters (cc) of fluids. What is the total intake of IV solution and fluids in liters?

12. Most adults have 5,000 to 6,000 milliliters of blood in their bodies. How many quarts of blood do they have? _____

13. Convert 225 pints to gallons. _____

14. How many liters are in 500 gallons? _____

15. A fire department pumps a total of 5,000 liters at a structure fire. How many gallons of water are pumped? _____

# Unit 28  MASS AND DENSITY MEASUREMENTS

## OBJECTIVE

Upon completion of this unit, the student should be able to

- calculate and/or convert weight, mass, and density measurement within and between the two measurement systems.

## BASIC PRINCIPLES OF MASS AND DENSITY MEASUREMENT

*Weight* is the measure of the downward force exerted on an object by the earth's gravity. In the English system, the standard unit of weight is the *pound*. *Mass* is the measure of the amount of matter in an object. The base unit of mass in the metric system is the *gram*. A constant used to convert between the English and metric systems is:

$$1 \text{ pound} = 454 \text{ grams}$$

**Example 1:**  Convert 25 pounds to grams.

**Solution:**
$$\frac{25 \text{ lb}}{1} \times \frac{454 \text{ g}}{1 \text{ lb}} = 11{,}350 \text{ g}$$

**Example 2:**  Convert 52,164 g to pounds.

**Solution:**
$$\frac{52{,}164 \text{ g}}{1} \times \frac{1 \text{ lb}}{454 \text{ g}} = \frac{52{,}164 \text{ lb}}{454} = 115 \text{ lb}$$

*Density* is the weight per unit volume or mass per unit volume of a substance. Several examples of density measurement are:

> grams per cubic centimeter ($g/cm^3$)
> pounds per cubic foot ($lb/ft^3$)
> pounds per gallon (lb/gal)

Important density measurements of water are:

> 1 cubic foot of water weighs 62.4 pounds for a density of 62.4 $lb/ft^3$
> 1 gallon of water weighs 8.34 pounds for a density of 8.34 lb/gal

The relationship between mass or weight (*m*), density (*D*), and volume (*V*) is given by:

$$D = \frac{m}{V}$$

Rearranging this formula allows you to solve for mass or volume:

$$m = VD \text{ or } V = \frac{m}{D}$$

**Example 1:** Calculate the weight of water in a 500-gallon on-board tank.

**Solution:** 1 gallon of water weighs 8.34 pounds. Solving for *m* gives

$$\begin{aligned} m &= VD \\ &= 500 \text{ gal} \times 8.34 \text{ lb/gal} \\ &= 4{,}170 \text{ lb} \end{aligned}$$

**Example 2:** Given that 1 cubic foot of water contains 7.48 gallons, show that the density of water expressed in lb/gal is 8.34 lb/gal.

**Solution:** Since 1 cubic foot of water weighs 62.4 pounds, solve for *D* as follows:

$$D = \frac{m}{V} = \frac{62.4 \text{ lb}}{7.48 \text{ gal}} = 8.34 \text{ lb/gal}$$

## PRACTICAL PROBLEMS

1. Convert 11,340 pounds to grams.

2. Convert 8,164.8 pounds to grams.

3. How many pounds do 500 grams weigh?

4. How many pounds do 249.48 grams weigh?

5. What is the weight of water in a 5,000-gallon tanker?

6. During a fire, approximately 3,750 gallons are pumped into a structure. What is the approximate weight of the water pumped into the structure?

7. What is the weight of water in a 50-cubic foot vessel?

8. What is the weight of a section of hose filled with water in which the hose measures 1¾ inches in diameter and 150 feet in length. (Hint: First convert diameter measurement to feet and find volume of cylindrical section of hose.)  _____

9. If a hose line is discharging 95 gallons per minute (gpm), what is the flow rate in pounds per minute? (Hint: 1 gallon of water weighs 8.34 pounds.)  _____

10. Find the volume of an object if its mass is 240 grams and its density is 10.5 g/cm$^3$.  _____

# Unit 29 TEMPERATURE MEASUREMENTS

**OBJECTIVE**

Upon completion of this unit, the student should be able to

- convert temperature measurements between Fahrenheit and Celsius.

**BASIC PRINCIPLES OF TEMPERATURE MEASUREMENT**

The most common temperature measurement used in the United States is the Fahrenheit scale (°F). Most other countries use the Celsius or Centigrade scale (°C). A comparison of the two systems is shown in the following diagram.

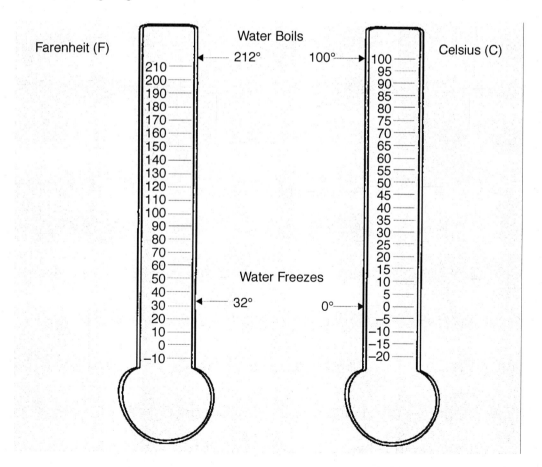

142  Section 5  Measurement

In the Fahrenheit system, the freezing point of water is 32° and the boiling point of water is 212°. In the Celsius system, the freezing point of water is 0° and the boiling point of water is 100°.

To convert Fahrenheit temperature to Celsius, the following formula is used:

$$C = (F - 32)5/9 \text{ or } C = (F - 32)0.556$$

**Example:** Convert 80° F to Celsius. Round answer to tenths.

**Solution:**
$$\begin{aligned} C &= (F - 32)5/9 \\ &= (80 - 32)5/9 \\ &= (48)(5/9) \\ &= 26\tfrac{2}{3} \text{ or } 26.7°C \end{aligned}$$
or
$$\begin{aligned} C &= (F - 32)0.556 \\ &= (80 - 32)0.556 \\ &= (48)(0.556) \\ &= 26.688 \text{ or } 26.7°C \end{aligned}$$

To convert Celsius temperature to Fahrenheit, the following formula is used:

$$F = 9/5\, C + 32 \qquad \text{or} \qquad F = 1.8\, C + 32$$

**Example:** Convert 60°C to Fahrenheit.

**Solution:**
$$\begin{aligned} F &= 9/5\, C + 32 \\ &= (9/5)(60) + 32 \\ &= 108 + 32 \\ &= 140°F \end{aligned}$$
or
$$\begin{aligned} F &= 1.8\, C + 32 \\ &= (1.8)(60) + 32 \\ &= 108 + 32 \\ &= 140°F \end{aligned}$$

## PRACTICAL PROBLEMS

Convert the following temperatures to Celsius degrees (°C). (Round to nearest unit.)

1. 59°F _____  4. 180°F _____

2. 131°F _____  5. 82.4°F _____

3. 40°F _____

Convert the following temperatures to Fahrenheit degrees (°F). (Round to nearest unit.)

6. 70°C _____  9. 48°C _____

7. 10°C _____  10. 18°C _____

8. 32°C _____

11. Normal oral body temperature is 37°C. What is normal oral body temperature in Fahrenheit to the nearest tenth? _____

12. The normal range for body temperature is 97°F to 100°F. What is the normal range for body temperature in Celsius to the nearest tenth? _____

13. The melting temperature of aluminum is 660.4°C. Convert the melting temperature to Fahrenheit. _____

14. The incipient, or first, stage of a fire is characterized as having ceiling temperatures up to 500°F. Convert this temperature to Celsius. _____

15. The flashpoint of gasoline is −45°F. What is the flashpoint of gasoline in Celsius? _____

# Statistics, Charts, and Graphs

**SECTION 6**

## Unit 30  INTRODUCTION TO STATISTICS

### OBJECTIVE

Upon completion of this unit, the student should be able to

- determine the mode, median, and mean for a set of data scores.

### BASIC PRINCIPLES OF STATISTICS

*Statistics* can be defined as a branch of mathematics that deals with the collection, analysis, interpretation, and presentation of numerical data. In this unit, the basic statistical measures of central tendency of data will be presented. In the following units, data in the form of charts and graphs will be presented. Charts and graphs provide a visual description or summary of data. (Note: *Data* is the plural form for *datum*, a single piece of information.)

### Central Tendency

*Central tendency* is one way to measure or describe the distribution of numerical values in a set of data. The most common measures of central tendency are the *mode*, *median*, and *mean*. Each of the three measures of central tendency provides information on how data are distributed. For example, an instructor can use central tendency measures for a set of test scores to determine class performance on a test.

### Mode

Mode is the simplest measure of central tendency and is defined as the most frequent value in the set. Mode is determined by simply looking at the data and identifying the number or score that appears most often. For example, in the following set of data, the mode is 250 gpm.

| Incident Number | Total GPM Pumped |
|---|---|
| 98-24 | 250 gpm |
| 98-59 | 90 gpm |
| 98-71 | 125 gpm |
| 98-74 | 300 gpm |
| 98-88 | 250 gpm |
| 98-102 | 200 gpm |
| 98-107 | 55 gpm |
| 98-125 | 250 gpm |
| 98-128 | 500 gpm |
| 98-130 | 200 gpm |

**Example:** Determine the mode for the following data:

$$25, 30, 12, 15, 30, 21, 15, 23, 40, 15$$

Note: It is possible for more than one mode to be present in a set of data.

**Solution:** 15 is the mode because it appears most frequently in the set (three times).

### Median

The *median* identifies the midpoint in the data. The median identifies a point above and below which an equal number of data lie. The first step in identifying the median is to arrange the scores in a set of data in ascending order (from lowest to highest). Next, if the number of scores is odd, the median is the middle score. Consider the following data:

$$3, 7, 9, 12, 20, 22, 25$$

Since the number of scores is odd (7 scores), the median is the middle, or fourth score, which is 12. If the number of scores is even, the median is halfway between the two middle scores. Consider the following data:

$$5, 7, 10, 11, 24, 25, 30, 33, 35, 40$$

Since the number of scores is even (10 scores), the median is halfway between 24 and 25, or 24.5.

**Example 1:** Determine the median for the following scores on an EMT exam.

$$75, 98, 82, 79, 95, 93, 80, 91, 89, 75, 100$$

**Solution:** First, arrange the data in ascending order:

$$75, 75, 79, 80, 82, 89, 91, 93, 95, 98, 100$$

Next, an odd number of scores indicates that the middle score is the median (89).

**Example 2:** Identify the median for the following data:

$$100, 55, 79, 200, 80, 90$$

**Solution:** Arrange data in ascending order:

$$55, 79, 80, 90, 100, 200$$

An even number of scores identifies the median as halfway between 80 and 90, or 85.

## Mean

The *arithmetic mean* (or simply, the *mean*) of a set of data is determined by adding all the scores and then dividing the sum by the total number of scores. Consider the data:

$$2, 10, 12, 20, 21$$

The mean is determined by dividing the sum of the scores (65) by the number of scores (5). The mean, then, is $65/5$ or 13.

**Example:** Calculate the mean for these Firefighter I certification exam results:

| Student | Grade |
|---|---|
| 1 | 90 |
| 2 | 73 |
| 3 | 85 |
| 4 | 99 |
| 5 | 80 |
| 6 | 79 |
| 7 | 95 |
| 8 | 90 |
| 9 | 65 |
| 10 | 86 |

**Solution:** The sum of the scores is 842. The number of scores is 10.

Therefore, the mean is $842/10 = 84.2$.

### CALCULATOR USE

Many scientific calculators have basic statistical capabilities. In most cases, the data are entered in a specific format and special commands are keyed in to conduct the statistical calculation. Be sure that you become familiar with the way your calculator enters data.

### PRACTICAL PROBLEMS

1. Determine the mode, median, and mean for the following data:
   10, 8, 25, 8, 11

   Mode: _____

   Median: _____

   Mean: _____

2. Determine the mode, median, and mean for the following data:
   25, 12, 15, 22, 19, 13, 12, 23

   Mode: _____

   Median: _____

   Mean: _____

3. Three firefighters challenge each other to a walking contest. The data for miles walked within a 7-day period is provided in the chart that follows. Determine the mode, median, and mean for each of the firefighters.

   FF1 Mode: _____

         Median: _____

         Mean: _____

   FF2 Mode: _____

         Median: _____

         Mean: _____

   FF3 Mode: _____

         Median: _____

         Mean: _____

| Firefighter 1 | | Firefighter 2 | | Firefighter 3 | |
| --- | --- | --- | --- | --- | --- |
| Day | Miles | Day | Miles | Day | Miles |
| 1 | 1 | 1 | 1 | 1 | 3 |
| 2 | 1.25 | 2 | 1 | 2 | 1 |
| 3 | .75 | 3 | .5 | 3 | 2 |
| 4 | 1 | 4 | 1 | 4 | 1 |
| 5 | 1.5 | 5 | .25 | 5 | .75 |
| 6 | 1 | 6 | 0 | 6 | .5 |
| 7 | 2 | 7 | 3 | 7 | 1.5 |

4. The results of a captain's promotional exam are listed in the following chart. Determine the mode, median, and the mean for the exam.

Mode: _____

Median: _____

Mean: _____

| \multicolumn{6}{c|}{Captain Promotional Exam Results} ||||||
|---|---|---|---|---|---|
| Student | Grade | Student | Grade | Student | Grade |
| 1 | 75 | 12 | 77 | 23 | 88 |
| 2 | 83 | 13 | 55 | 24 | 97 |
| 3 | 66 | 14 | 97 | 25 | 82 |
| 4 | 91 | 15 | 67 | 26 | 69 |
| 5 | 75 | 16 | 79 | 27 | 80 |
| 6 | 90 | 17 | 80 | 28 | 77 |
| 7 | 96 | 18 | 99 | 29 | 75 |
| 8 | 77 | 19 | 78 | 30 | 80 |
| 9 | 86 | 20 | 69 | 31 | 82 |
| 10 | 80 | 21 | 97 | 32 | 97 |
| 11 | 79 | 22 | 90 | 33 | 80 |

5. The response times for an ambulance are: 4.5 minutes, 3 minutes, 2.75 minutes, 7 minutes, 5 minutes, 6 minutes, and 3 minutes. Determine the mode, median, and mean.

Mode: _____

Median: _____

Mean: _____

# Unit 31   LINE GRAPHS

## OBJECTIVE

Upon completion of this unit, the student should be able to

- develop and interpret line graphs.

## BASIC PRINCIPLES OF LINE GRAPHS

*Line graphs* are used to illustrate the behavior of the physical quantity represented by one variable compared to that of another variable. One variable is placed on the x-axis (horizontal axis) and the second variable is placed on the y-axis (vertical axis). Often, the variable for time is placed on the horizontal axis and the variable for the quantity measured over time is placed on the vertical axis. Line graphs provide an excellent visual description of data. In some cases, line graphs can also be used to predict values.

**Example:**   From the line graph presented here, determine the following:

   a. Temperature after 5 minutes.

   b. Temperature after 30 minutes.

   c. Time when temperature reaches 1300°F.

   d. What does the line graph tell us about temperature within the first 5 to 10 minutes of a fire?

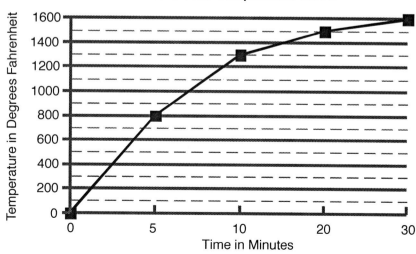

**Solution:**

a. 800°F

b. 1600°F

c. 10 minutes

d. There is a very rapid rise in temperature within the first 5 to 10 minutes of a fire.

**PRACTICAL PROBLEMS**

For each of the line graphs, provide the requested information.

1. Use the graph on firefighter fatalities while responding to or returning from an incident to answer these questions.

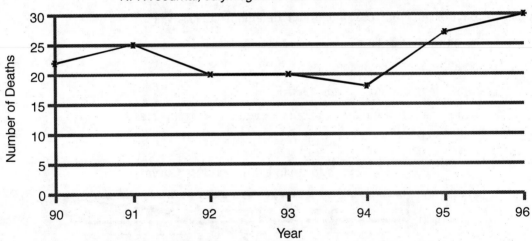

a. How many fatalities occurred during 1992 and 1993? _____
b. During what year did more fatalities occur than in any other year? _____
c. What year had the fewest fatalities? _____
d. What is the trend for fatalities over the last few years shown? _____

2. Use the graph comparing career and volunteer firefighter deaths to answer these questions.

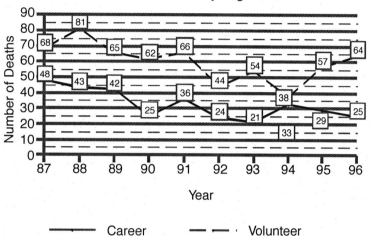

a. How many career firefighter fatalities occurred in 1991? _____

b. Which year had the most volunteer fatalities? _____

c. Which year had the combined highest number of fatalities? _____

d. What is the trend based on the last three years shown for career and volunteer firefighter fatalities? _____

3. Use the quarterly run report graph showing totals for three ambulances to answer these questions.

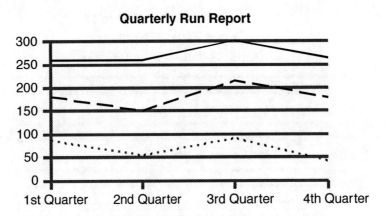

a. What variable is on the y-axis and what variable is on the x-axis?

y-axis: _____

x-axis: _____

b. Which ambulance consistently has more runs? _____

c. Which quarter has the fewest runs for each of the ambulances? _____

d. What trends can be determined from the line graph? _____

4. Plot the following data on the chart provided. Connect the data points to create a line chart.

| Year | Civilian Home Fire Deaths |
|------|---------------------------|
| 1990 | 4,050 |
| 1991 | 3,500 |
| 1992 | 3,705 |
| 1993 | 3,720 |
| 1994 | 3,425 |
| 1995 | 3,640 |
| 1996 | 4,035 |

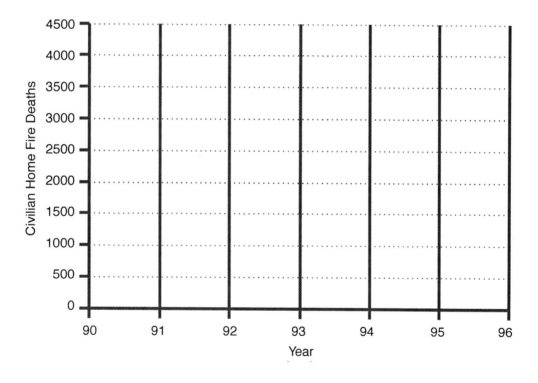

5. The following line graph shows the velocity of a moving object versus time in seconds. Using the graph, estimate the speed of the object at 4, 6, 11, and 16 seconds.

4 seconds: _____

6 seconds: _____

11 seconds: _____

16 seconds: _____

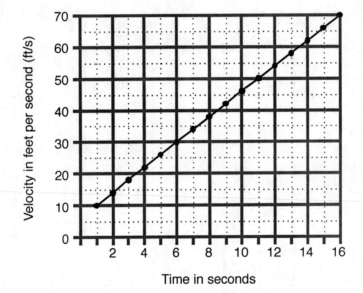

# Unit 32  BAR GRAPHS

**OBJECTIVE**

Upon completion of this unit, the student should be able to

- develop and interpret bar graphs.

**BASIC PRINCIPLES OF BAR GRAPHS**

*Bar graphs* are used to compare the numerical values of several variables. The height or length of the bars indicates the relationship between the variables. The bars in the graph can be vertical, horizontal, and/or stacked but must always be of the same width. Several examples of bar graphs follow.

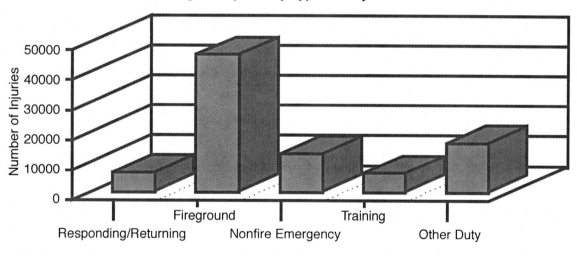

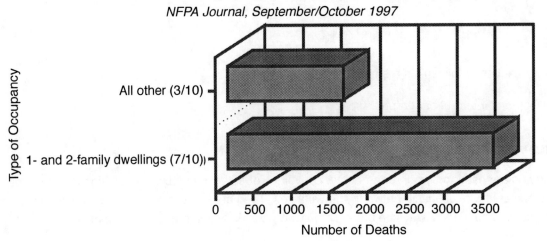

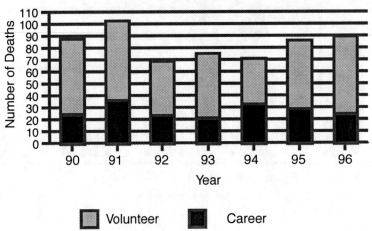

## PRACTICAL PROBLEMS

1. The following bar graph shows the amount of fiber in various foods.

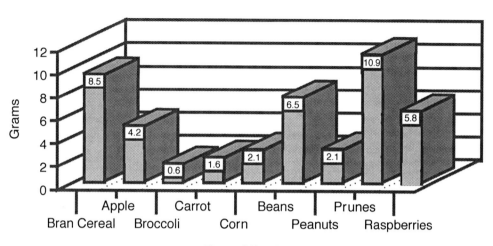

a. Which three foods contain the highest fiber content? _____

b. Which four foods contain the lowest fiber content? _____

c. How many food types contain over 3 grams of fiber? _____

d. Which types of food have 2.1 grams of fiber? _____

2. The bar graph that follows presents the data from a three month analysis on ambulance runs for the north and south sides of a city.

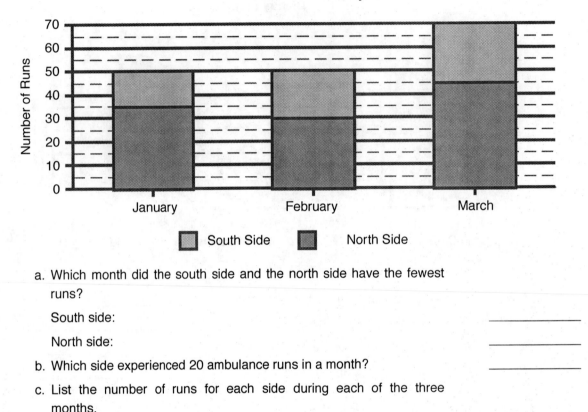

a. Which month did the south side and the north side have the fewest runs?

South side: _____

North side: _____

b. Which side experienced 20 ambulance runs in a month? _____

c. List the number of runs for each side during each of the three months.

South side — January: _____

February: _____

March: _____

North side — January: _____

February: _____

March: _____

3. The following bar graph presents the data on deaths associated with fireworks accidents from 1983 to 1993.

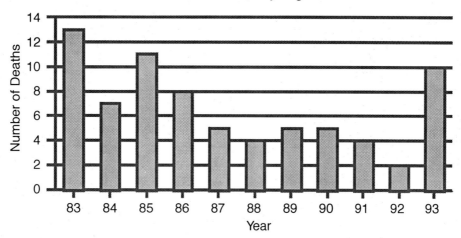

**Deaths Associated with Fireworks Accidents**
*NFPA Journal, July/August 1997*

a. List the three years with the highest number of deaths associated with fireworks. _____

b. How many deaths occurred in 1987, 1991, and 1993? _____

c. Which three years experienced the fewest deaths associated with fireworks? _____

d. How would you describe the data between 1983 and 1992? What about from 1983 to 1993?

1983 to 1992: _____

1983 to 1993: _____

4. The next bar graph presents data of on-duty firefighter deaths by age and cause of death during 1996.

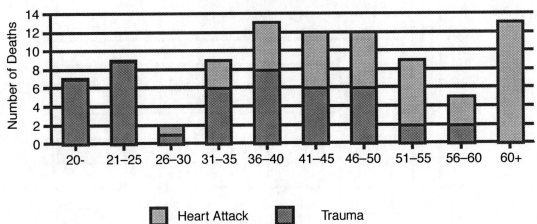

a. From looking at the bar graph, what happens to cause of death (trauma versus heart attack) as age increases? _____

b. Which caused more deaths in 1996, trauma or heart attack?
_____

c. How many firefighter deaths were caused by trauma for the 31 to 35 age group? _____

d. Which age group had the fewest deaths from either heart attack or trauma? _____

5. Given the following data, finish the bar graph that has been started.

| Week | Number of Runs |
|------|----------------|
| 1    | 13             |
| 2    | 8              |
| 3    | 12             |
| 4    | 9              |

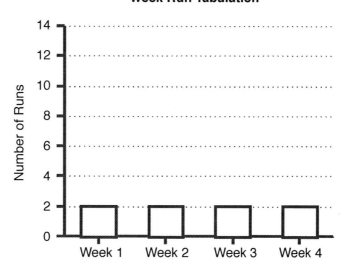

# Unit 33  PIE CHARTS

**OBJECTIVE**

Upon completion of this unit, the student should be able to

- interpret pie charts.

**BASIC PRINCIPLES OF PIE CHARTS**

*Pie charts* are typically used when data are in percentage form. The chart is divided into wedges, each representing a percentage of the total. The entire circle on a pie chart represents 100 percent. The following pie chart is an example of this division.

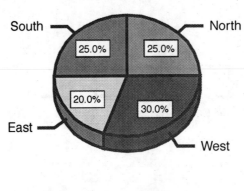

**PRACTICAL PROBLEMS**

1. Using the pie chart of emergency room admissions, answer these questions.

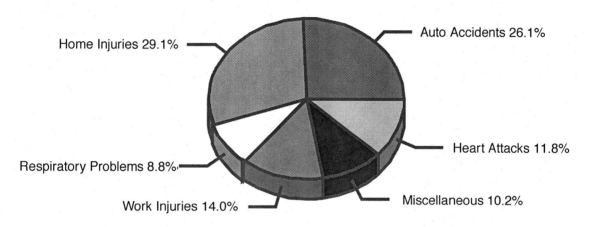

a. Which two categories combined represent over 50 percent of emergency room admissions? _____

b. Which category represents the smallest percentage of admissions? _____

c. What percent of admissions is the result of work injuries? _____

d. What is the percent of admissions resulting from heart attacks? _____

2. Mark the following pie chart to identify the data presented here.

**Firefighter Fatalities by Rank, 1996**
(Total of 92 Fatalities)

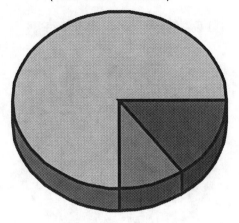

a. Company officers represent 10 percent of all on-duty firefighter fatalities during 1996.

b. Firefighters represent 75 percent of all on-duty firefighter fatalities during 1996.

c. Chief officers represent 15 percent of all on-duty firefighter fatalities during 1996.

3. Label the wedges of the pie chart on firefighter deaths by cause of injury for 1996 based on these data:

| | |
|---|---|
| Caught/Trapped | 8.7% |
| Stress | 50% |
| Exposure | 7.6% |
| Struck by/Contact with Object | 32.6% |
| Fell | 1.1% |

**Firefighter Deaths by Cause of Injury, 1996**
*NFPA Journal, July/August 1997*

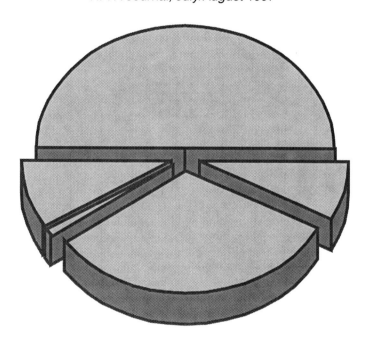

Which two categories make up over 80% of causes of injuries?

4. Use the pie chart of blood types to answer the following questions.

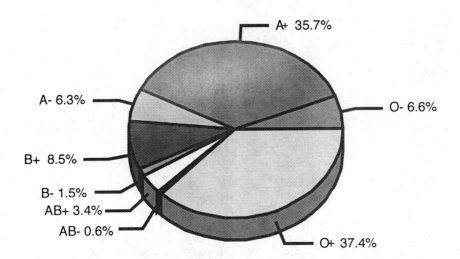

**Blood Types**

- A+ 35.7%
- O- 6.6%
- O+ 37.4%
- AB- 0.6%
- AB+ 3.4%
- B- 1.5%
- B+ 8.5%
- A- 6.3%

a. Which two blood types are the most common?  _____

b. Which is the least common blood type?  _____

c. What are the percentages for the following blood types: A– and B–?  _____

5. Use the pie chart presenting data on fireworks injuries by body part for 1996 to answer these questions.

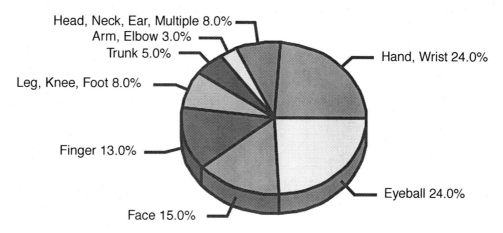

**Fireworks Injuries by Body Part, 1996**

- Head, Neck, Ear, Multiple 8.0%
- Arm, Elbow 3.0%
- Trunk 5.0%
- Leg, Knee, Foot 8.0%
- Finger 13.0%
- Face 15.0%
- Hand, Wrist 24.0%
- Eyeball 24.0%

a. Which two categories represent almost 50 percent of body parts injured by fireworks?

b. What are the top three body parts injured by fireworks in 1996?

c. What body parts are the least likely to be injured by fireworks?

d. Wearing proper safety gloves and goggles can help prevent what percentage of body parts typically injured by fireworks?

# Formulas and Equations

SECTION 7

## Unit 34  INTRODUCTION TO FORMULAS AND EQUATIONS

### OBJECTIVE

Upon completion of this unit, the student should be able to

- develop formulas and equations from narratives.

### BASIC PRINCIPLES OF FORMULAS

A *formula* is a mathematical statement or *equation* expressing a relationship between certain physical quantities. Several formulas, such as temperature conversion and needed flow, were introduced in earlier units. Standard formulas are typically obtained from reference materials, such as the appendices of textbooks, trade handbooks, or technical manuals published by trade or professional associations.

Formulas contain constants, variables, and mathematical operation symbols. *Constants* are numbers or values that do not change, such as pi ($\pi$). *Variables* represent physical quantities that change in value under different situations, such as gpm (gallons per minute) or psi (pounds per square inch). *Mathematical operation symbols* indicate the mathematical procedures to perform in the formula, such as multiplication or subtraction. Variable symbols indicate the type of variable, and sometimes units to use in the formula.

A practical problem describes a problem in a *narrative format*, so the problem must be translated into variables and mathematical symbols. Many of the terms used in practical problems and their associated symbols were introduced in earlier units and should be reviewed as needed. When expressing a practical problem as an equation, it is important to identify all variables, units, constants, and mathematical operations. The variables and any constants are combined with the appropriate mathematical symbols to form the equation. It may be helpful to draw a diagram to describe the relationships in the problem.

**Example:**  Write the following practical problem as an equation:

The needed flow ($NF$) for a fire is equal to the volume ($V$) of the structure in cubic feet divided by the constant 100 ft$^3$/gpm.

*172 Section 7 Formulas and Equations*

**Solution:** $NF = V/100 \text{ ft}^3$
where $NF$ = Needed Flow
$V$ = Volume in cubic feet
$100 \text{ ft}^3/\text{gpm}$ = Constant

## PRACTICAL PROBLEMS

For each of the problems, write the mathematical formula or equation. Use the symbols stated or implied in the problem.

1. The radius ($r$) of a circle is equal to one-half times the diameter ($d$).  _____

2. A temperature measurement in Fahrenheit degrees (°F) is equal to $9/5$ multiplied by the Celsius reading (°C), plus 32 degrees.  _____

3. The volume ($V$) of a rectangle is equal to the length ($L$) multiplied by the width ($W$) multiplied by the height ($H$).  _____

4. The total amount ($A$) repaid for a loan is equal to the principal ($P$) plus the interest ($I$).  _____

5. The needed flow ($NF$) is equal to the area ($A$) in square feet divided by the constant 3 ft²/gpm.  _____

6. The pressure gain or loss caused by elevation ($EL$) is equal to the difference of height ($H$) in feet multiplied by the constant 0.5 psi/ft.  _____

7. Pressure ($P$) is equal to the force ($F$) exerted by a substance divided by the surface area ($A$) over which the force is exerted.  _____

8. The weight ($W$) of a substance is equal to its density ($d$) multiplied by its volume ($V$).  _____

9. The pump discharge pressure ($PDP$) is equal to the friction loss ($FL$) added to the nozzle pressure ($NP$) added to the gain or loss in pressure caused by elevation ($EL$) added to any appliance friction loss ($AFL$) in the hose line.  _____

10. Nozzle reaction ($NR$) is equal to the nozzle diameter squared ($d^2$) multiplied by the nozzle pressure ($NP$) multiplied by the constant 1.57.  _____

# Unit 35  COMMON EMERGENCY SERVICE FORMULAS AND EQUATIONS

## OBJECTIVE

Upon completion of this unit, the student should be able to

- use common emergency service formulas and equations to solve practical problems.

## BASIC PRINCIPLES FOR SOLVING FORMULAS AND EQUATIONS

After a formula or equation has been developed or identified, the process for solving a problem is to simply plug in the known values and complete the operations in the appropriate order.

**Example:** Calculate the needed flow (*NF*) for a one-story structure measuring 60 feet wide by 120 feet long by 10 feet high.

Use the Iowa formula: $NF = \dfrac{V}{100 \text{ ft}^3/\text{gpm}}$

where  $NF$ = Needed Flow
$V$ = Volume of the area in cubic feet ($ft^3$)
$100 \ (ft^3/gpm)$ = a constant in cubic feet per gallons per minute

**Solution:** First determine the volume of the area:

$$60 \text{ ft} \times 120 \text{ ft} \times 10 \text{ ft} = 72{,}000 \text{ ft}^3$$

Next, plug the numbers into the formula:

$$NF = \dfrac{72{,}000 \text{ ft}^3}{100 \text{ ft}^3/\text{gpm}} = \dfrac{720}{1 \text{ gpm}} = 720 \text{ gpm}$$

Several helpful hints when solving formulas include:

- use the right formula/properly translate the problem into an equation
- make sure you are working with proper units (convert as necessary)
- keep units straight
- write down the answer to each step to avoid confusion and to troubleshoot wrong answers
- follow the proper order of operations

## Section 7  Formulas and Equations

**PRACTICAL PROBLEMS**

1. Use these nozzle reaction (NR) formulas to answer the following questions.

   Smooth Bore Nozzle
   $NR = 1.57/in^3 \times d^2 \times NP$
   where
   - $NR$ = nozzle reaction in pounds
   - $1.57/in^3$ = constant
   - $d$ = diameter of nozzle orifice in inches
   - $NP$ = nozzle pressure in psi

   Combination Nozzle
   $NR = gpm \times \sqrt{NP} \times 0.0505\ in^2/gpm$
   where
   - $NR$ = nozzle reaction in pounds
   - $gpm$ = gallons per minute
   - $NP$ = nozzle pressure
   - $0.0505\ in^2/gpm$ = constant

   a. What is the nozzle reaction for a ⅜-inch diameter smooth bore nozzle with a nozzle pressure of 50 psi? _____

   b. What nozzle reaction will be caused by a combination nozzle flowing 95 gpm with a nozzle pressure of 100 psi? _____

2. Use these needed flow (NF) formulas to answer the following question.

   Iowa Formula
   $NF = \dfrac{V}{100}$
   where
   - $NF$ = needed flow in gpm
   - $V$ = volume of the structure in $ft^3$
   - $100\ ft^3/gpm$ = constant

   National Fire Academy (NFA)
   $NF = \dfrac{A}{3}$
   where
   - $NF$ = needed flow in gpm
   - $A$ = area of a structure in $ft^2$
   - $3\ ft^2/gpm$ = constant

   What is the needed flow of water for a structure measuring 30 feet wide by 60 feet long with a 9-foot high ceiling? Use both the Iowa and the NFA formulas. (Note: The two formulas are based on different assumptions about a fire and related fire-extinguishing tactics. Therefore, the formulas provide different results.) _____

3. Use this intravenous flow rate formula to answer the following question.

   $FR = V \times DF \times T$
   where
   
   $FR$ = flow rate in drops per min (gtt/min)
   $V$ = volume over time administered in ml/hr
   $DF$ = drop factor in gtt/ml
   $T$ = Time

   A physician orders 1,000 ml of 5% dextrose in water ($D_5W$) to be infused in 8 hours. Volume ($V$) is equal to 1,000 ml/8 hr. The drop factor ($DF$) of the infusion set is calibrated at 20 gtt/ml. What is the flow rate ($FR$) in drops per minute?

   Note: The formula can be rewritten in terms of units used as:

   $$FR = \frac{ml}{hr} \times \frac{ggt}{ml} \times \frac{1\ hr}{60\ min}$$

176   Section 7   Formulas and Equations

4. When a hose dispenses water to a destination above the water pump, there is a loss in water pressure. When a hose dispenses water to a destination below the water pump, there is a gain in water pressure. The elevation gain or loss is estimated by the elevation gain or loss formula:

$EL = 0.5 \text{ psi/ft} \times H$
where
   $EL$ = gain or loss of elevation in psi
   0.5 psi/ft = constant
   $H$ = height (difference in elevation) in feet

   a. A hose line is operating 50 feet above the pump. What is the estimated pressure loss caused by elevation? _____

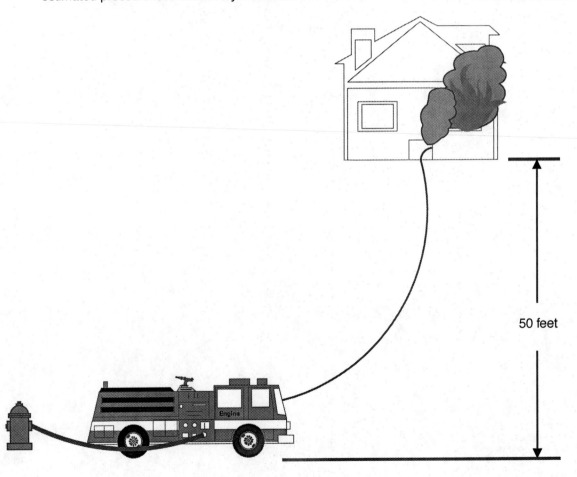

b. A hose line is operating 30 feet below the pump. What is the estimated gain in pressure caused by elevation? (Hint: The answer will be a negative number.) _____

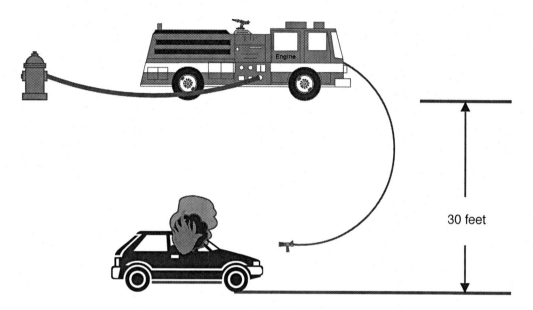

5. Use this friction loss formula to answer the following questions.

$FL = c \times q^2 \times L$

where

    $FL$ = friction loss in the hose in psi

    $c$ = constant for a specific hose diameter in psi/gpm²ft
       (24 psi/gpm²ft for 1½-inch hose; 15.5 psi/gpm²ft for 1¾-inch hose)

    $q$ = flow in hundreds of gallons per minute $\frac{gpm}{100}$

    $L$ = length of hose in hundreds of feet ($\frac{ft}{100}$)

Note: The units cancel as follows:

$$FL = \frac{\cancel{psi}}{\cancel{gpm^2 \, ft}} \times \frac{\cancel{gpm^2}}{100^2} \times \frac{\cancel{ft}}{100}$$

178   Section 7   Formulas and Equations

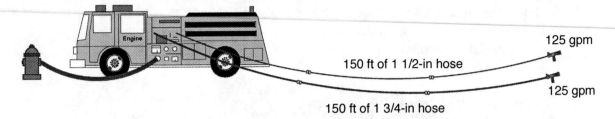

    a. What is the friction loss in 150 feet of 1½-inch hose flowing 125 gpm?  _____

    b. What is the friction loss in 150 feet of 1¾-inch hose flowing 125 gpm?  _____

6. Use the following pump discharge pressure formula to solve this problem.

   PDP = NP + FL + AFL + EL
   where

       PDP = pump discharge pressure in psi
       NP = nozzle operating pressure in psi
       FL = friction loss in the hose in psi
       AFL = appliance friction loss in psi
       EL = elevation gain or loss in psi

   Determine the pump discharge pressure (PDP) when a hose line has 30 psi of friction loss (FL), 5 psi of appliance friction loss (AFL), −15 psi of elevation loss (EL), and is operating with 100 psi of nozzle pressure (NP).  _____

7. Use the following head (height) formula to solve this problem.

   h = 2.31 ft/psi × P
   where

       h = height of water in feet
       2.31 ft/psi = height in feet that 1 psi will raise water (constant)
       P = pressure in psi

   Calculate the head (height) of water if the pressure is 43.3 psi.  _____

8. Use this pressure formula to answer the following question.

   $P = 0.433 \text{ lb/in}^2\text{ft} \times h$
   where
   > $P$ = pressure in psi
   > $0.433$ = constant in lb/in²ft
   > $h$ = height (depth) of the water in feet

   If a vessel filled with water is 200 feet high, what is the pressure at the bottom of the vessel? _____

9. Use this pediatric dosage formula (Fried's Rule for children under 2 years of age) to answer the following question.

   $$\text{Infant Dose} = \frac{\text{Age in months}}{150 \text{ months}} \times \text{Adult Dose}$$

   The usual adult dose for Demerol is 50 milligrams (mg). An infant is 10 months old. What is the correct dosage? _____

10. Use the following IV infusion time formula to solve this problem.

    $$IF = \frac{\text{Total volume to infuse}}{\text{ml/hour being infused}}$$

    where
    > $IF$ = infusion time in hours

    Calculate the infusion time for an IV of 1,000 ml of 5% dextrose in water ($D_5W$) infusing at 50 ml/hour. _____

# Appendix

## Section 1: ENGLISH RELATIONSHIPS

### ENGLISH LENGTH MEASURE

| | | |
|---|---|---|
| 1 foot (ft) | = | 12 inches (in) |
| 1 yard (yd) | = | 3 feet (ft) |
| 1 mile (mi) | = | 1,760 yard (yd) |
| 1 mile (mi) | = | 5,280 feet (ft) |

### ENGLISH AREA MEASURE

| | | |
|---|---|---|
| 1 square yard (sq yd) | = | 9 square feet (sq ft) |
| 1 square foot (sq ft) | = | 144 square inches (sq in) |
| 1 square mile (sq mi) | = | 640 acres |
| 1 acre | = | 43,560 square feet (sq ft) |

### ENGLISH VOLUME MEASURE FOR SOLIDS

| | | |
|---|---|---|
| 1 cubic yard (cu yd) | = | 27 cubic feet (cu ft) |
| 1 cubic foot (cu ft) | = | 1,728 cubic inches (cu in) |

### ENGLISH VOLUME MEASURE FOR FLUIDS

| | | |
|---|---|---|
| 1 quart (qt) | = | 2 pints (pt) |
| 1 gallon (gal) | = | 4 quarts (qt) |

### ENGLISH VOLUME MEASURE EQUIVALENTS

| | | |
|---|---|---|
| 1 gallon (gal) | = | 0.133681 cubic foot (cu ft) |
| 1 gallon (gal) | = | 231 cubic inches (cu in) |

## Section 2: METRIC RELATIONSHIPS

The base units in SI metrics include the meter and the gram. Other units of measure are related to these units. The relationship between the units is based on powers of ten and uses these prefixes:

kilo (1,000)    hecto (100)    deka (10)    deci (0.1)    centi (0.01)    milli (0.001)

These tables show the most frequently used units with an asterisk (*)

### METRIC LENGTH

| | | |
|---|---|---|
| 10 millimeters (mm)* | = | 1 centimeter (cm)* |
| 10 centimeters (cm) | = | 1 decimeter (dm) |
| 10 decimeters (dm) | = | 1 meter (m)* |
| 10 meters (m) | = | 1 dekameter (dam) |
| 10 dekameters (dam) | = | 1 hectometer (hm) |
| 10 hectometers (hm) | = | 1 kilometer (km)* |

- To express a metric length unit as a smaller metric length unit, multiply by a positive power of ten such as 10, 100, 1,000, 10,000, and so on.
- To express a metric length unit as a larger metric length unit, multiply by a negative power of ten such as 0.1, 0.01, 0.001, 0.0001, and so on.

### METRIC AREA MEASURE

| | | |
|---|---|---|
| 100 square millimeters ($mm^2$) | = | 1 square centimeter ($cm^2$)* |
| 100 square centimeters ($cm^2$) | = | 1 square decimeter ($dm^2$) |
| 100 square decimeters ($cm^2$) | = | 1 square meter ($m^2$)* |
| 100 square meters ($m^2$) | = | 1 square dekameter ($dam^2$) |
| 100 square dekameters ($dam^2$) | = | 1 square hectometer ($hm^2$)* |
| 100 square hectometers ($hm^2$) | = | 1 square kilometer ($km^2$) |

- To express a metric area unit as a smaller metric area unit, multiply by 100, 10,000, 1,000,000, and so on.
- To express a metric area unit as a larger metric area unit, multiply by 0.01, 0.0001, 0.000001, and so on.

### METRIC VOLUME MEASURE FOR SOLIDS

| | | |
|---|---|---|
| 1,000 cubic millimeters ($mm^3$) | = | 1 cubic centimeter (cm)* |
| 1,000 cubic centimeters ($cm^3$) | = | 1 cubic decimeter ($dm^3$)* |
| 1,000 cubic decimeters ($dm^3$) | = | 1 cubic meter ($m^3$)* |
| 1,000 cubic meters ($m^3$) | = | 1 cubic dekameter ($dam^3$) |
| 1,000 cubic dekameters ($dam^3$) | = | 1 cubic hectometer ($hm^3$) |
| 1,000 cubic hectometers ($hm^3$) | = | 1 cubic kilometer ($km^3$) |

- To express a metric volume unit for solids as a smaller metric volume unit for solids, multiply by 1,000, 1,000,000, 1,000,000,000, and so on.
- To express a metric volume unit for solids as a larger metric volume unit for solids, multiply by 0.001, 0.000001, 0.000000001, and so on.

### METRIC VOLUME MEASURE FOR FLUIDS

| | | |
|---|---|---|
| 10 milliliters (mL)* | = | 1 centiliter (cL) |
| 10 centiliters (cL) | = | 1 deciliter (dL) |
| 10 deciliters (dL) | = | 1 liter (L)* |
| 10 liters (L) | = | 1 dekaliter (daL) |
| 10 dekaliters (daL) | = | 1 hectoliter (hL) |
| 10 hectoliters (hL) | = | 1 kiloliter (kL) |

- To express a metric volume unit for fluids as a smaller metric volume unit for fluids, multiply by 10, 100, 1,000, 10,000, and so on.
- To express a metric volume unit for fluids as a larger metric volume unit for fluids, multiply by 0.1, 0.01, 0.001, 0.0001, and so on.

## METRIC VOLUME MEASURE EQUIVALENTS

| | |
|---|---|
| 1 cubic decimeter (dm³) | = 1 liter (L) |
| 1,000 cubic centimeters (cm³) | = 1 liter (L) |
| 1 cubic centimeter (cm³) | = 1 milliliter (mL) |

## METRIC MASS MEASURE

| | |
|---|---|
| 10 milligrams (mg)* | = 1 centigram (cg) |
| 10 centigrams (cg) | = 1 decigram (dg) |
| 10 decigrams (dg) | = 1 gram (g)* |
| 10 grams (g) | = 1 dekagram (dag) |
| 10 dekagrams (dag) | = 1 hectogram (hg) |
| 10 hectograms (hg) | = 1 kilogram (kg)* |
| 1,000 kilograms (kg) | = 1 megagram (Mg)* |

◆ To express a metric mass unit as a smaller metric mass unit, multiply by 10, 100, 1,000, 10,000, and so on.
◆ To express a metric mass unit as a larger metric mass unit, multiply by 0.1. 0.01, 0.001, and so on.

Metric measurements are expressed in decimal parts of a whole number. For example, one-half millimeter is written as 0.5 mm.

In calculating with the metric system, all measurements are expressed using the same prefixes. If answers are needed in millimeters, all parts of the problem should be expressed in millimeters before the final solution is attempted. Diagrams that have dimensions in different prefixes must first be expressed using the same unit.

## Section 3: ENGLISH-METRIC EQUIVALENTS

### LENGTH MEASURE

| | | |
|---:|:---:|:---|
| 1 inch (in) | = | 25.4 millimeters (mm) |
| 1 inch (in) | = | 2.54 centimeters (cm) |
| 1 foot (ft) | = | 0.3048 meter (m) |
| 1 yard (yd) | = | 0.9144 meter (m) |
| 1 mile (mi) | = | 1.609 kilometers (km) |
| 1 millimeter (mm) | = | 0.03937 inch (in) |
| 1 centimeter (cm) | = | 0.39370 inch (in) |
| 1 meter (m) | = | 3.28084 feet (ft) |
| 1 meter (m) | = | 1.09361 yards (yd) |
| 1 kilometer (km) | = | 0.62137 mile (mi) |

### AREA MEASURE

| | | |
|---:|:---:|:---|
| 1 square inch (sq in) | = | 645.16 square millimeters ($mm^2$) |
| 1 square inch (sq in) | = | 6.4516 square centimeters ($cm^2$) |
| 1 square foot (sq ft) | = | 0.092903 square meter ($m^2$) |
| 1 square yard (sq yd) | = | 0.836127 square meter ($m^2$) |
| 1 square millimeter ($mm^2$) | = | 0.001550 square inch (sq in) |
| 1 square centimeter ($cm^2$) | = | 0.15500 square inch (sq in) |
| 1 square meter ($m^2$) | = | 10.763910 square feet (sq ft) |
| 1 square meter ($m^2$) | = | 1.19599 square yards (sq yd) |

### VOLUME MEASURE FOR SOLIDS

| | | |
|---:|:---:|:---|
| 1 cubic inch (cu in) | = | 16.387064 cubic centimeters ($cm^3$) |
| 1 cubic foot (cu ft) | = | 0.028317 cubic meter ($m^3$) |
| 1 cubic yard (cu yd) | = | 0.764555 cubic meter ($m^3$) |
| 1 cubic centimeter ($cm^3$) | = | 0.061024 cubic inch (cu in) |
| 1 cubic meter ($m^3$) | = | 35.314667 cubic feet (cu ft) |
| 1 cubic meter ($m^3$) | = | 1.307951 cubic yards (cu yd) |

### VOLUME MEASURE FOR FLUIDS

| | | |
|---:|:---:|:---|
| 1 gallon (gal) | = | 3,785.411 cubic centimeters ($cm^3$) |
| 1 gallon (gal) | = | 3.785411 liters (L) |
| 1 quart (qt) | = | 0.946353 liter (L) |
| 1 ounce (oz) | = | 29.573530 cubic centimeters ($cm^3$) |
| 1 cubic centimeter ($cm^3$) | = | 0.000264 gallon (gal) |
| 1 liter (L) | = | 0.264172 gallon (gal) |
| 1 liter (L) | = | 1.056688 quarts (qt) |
| 1 cubic centimeter ($cm^3$) | = | 0.033814 ounce (oz) |

### MASS MEASURE

| | | |
|---:|:---:|:---|
| 1 pound (lb) | = | 0.453592 kilogram (kg) |
| 1 pound (lb) | = | 453.59237 grams (g) |
| 1 ounce (oz) | = | 28.349523 grams (g) |
| 1 ounce (oz) | = | 0.028350 kilogram (kg) |
| 1 kilogram (kg) | = | 2.204623 pounds (lb) |
| 1 gram (g) | = | 0.002205 pound (lb) |
| 1 kilogram (kg) | = | 35.273962 ounces (oz) |
| 1 gram (g) | = | 0.035274 ounce (oz) |

# Section 4: FRACTION/DECIMAL EQUIVALENTS

| Fraction | Decimal Equivalent Customary (in) | Metric (mm) | Fraction | Decimal Equivalent Customary (in) | Metric (mm) |
|---|---|---|---|---|---|
| 1/64 | .015625 | 0.3969 | 33/64 | .515625 | 13.0969 |
| 1/32 | .03125 | 0.7938 | 17/32 | .53125 | 13.4938 |
| 3/64 | .046875 | 1.1906 | 35/64 | .546875 | 13.8906 |
| 1/16 | .0625 | 1.5875 | 9/16 | .5625 | 14.2875 |
| 5/64 | .078125 | 1.9844 | 37/64 | .578125 | 14.6844 |
| 3/32 | .09375 | 2.3813 | 19/32 | .59375 | 15.0813 |
| 7/64 | .109375 | 2.7781 | 39/64 | .609375 | 15.4781 |
| 1/8 | .1250 | 3.1750 | 5/8 | .6250 | 15.8750 |
| 9/64 | .140625 | 3.5719 | 41/64 | .640625 | 16.2719 |
| 5/32 | .15625 | 3.9688 | 21/32 | .65625 | 16.6688 |
| 11/64 | .171875 | 4.3656 | 43/64 | .671875 | 17.0656 |
| 3/16 | .1875 | 4.7625 | 11/16 | .6875 | 17.4625 |
| 13/64 | .203125 | 5.1594 | 45/64 | .703125 | 17.8594 |
| 7/32 | .21875 | 5.5563 | 23/32 | .71875 | 18.2563 |
| 15/64 | .234375 | 5.9531 | 47/64 | .734375 | 18.6531 |
| 1/4 | .250 | 6.3500 | 3/4 | .750 | 19.0500 |
| 17/64 | .265625 | 6.7469 | 49/64 | .765625 | 19.4469 |
| 9/32 | .28125 | 7.1438 | 25/32 | .78125 | 19.8438 |
| 19/64 | .296875 | 7.5406 | 51/64 | .796875 | 20.2406 |
| 5/16 | .3125 | 7.9375 | 13/16 | .8125 | 20.6375 |
| 21/64 | .328125 | 8.3384 | 53/64 | .828125 | 21.0344 |
| 11/32 | .34375 | 8.7313 | 27/32 | .84375 | 21.4313 |
| 23/64 | .359375 | 9.1281 | 55/64 | .859375 | 21.8281 |
| 3/8 | .3750 | 9.5250 | 7/8 | .8750 | 22.2250 |
| 25/64 | .390625 | 9.9219 | 57/64 | .890625 | 22.6219 |
| 13/32 | .40625 | 10.3188 | 29/32 | .90625 | 23.0188 |
| 27/64 | .421875 | 10.7156 | 59/64 | .921875 | 23.4156 |
| 7/16 | .4375 | 11.1125 | 15/16 | .9375 | 23.8125 |
| 29/64 | .453125 | 11.5094 | 61/64 | .953125 | 24.2094 |
| 15/32 | .46875 | 11.9063 | 31/32 | .96875 | 24.6063 |
| 31/64 | .484375 | 12.3031 | 63/64 | .984375 | 25.0031 |
| 1/2 | .500 | 12.7000 | 1 | 1.000 | 25.4000 |

# Section 5: POWERS AND ROOTS OF NUMBERS (1 through 100)

| Number | Powers Square | Powers Cube | Roots Square | Roots Cube | Number | Powers Square | Powers Cube | Roots Square | Roots Cube |
|---|---|---|---|---|---|---|---|---|---|
| 1 | 1 | 1 | 1.000 | 1.000 | 51 | 2,601 | 132,651 | 7.141 | 3.708 |
| 2 | 4 | 8 | 1.414 | 1.260 | 52 | 2,704 | 140,608 | 7.211 | 3.733 |
| 3 | 9 | 27 | 1.732 | 1.442 | 53 | 2,809 | 148,877 | 7.280 | 3.756 |
| 4 | 16 | 64 | 2.000 | 1.587 | 54 | 2,916 | 157,464 | 7.348 | 3.780 |
| 5 | 25 | 125 | 2.236 | 1.710 | 55 | 3,025 | 166,375 | 7.416 | 3.803 |
| 6 | 36 | 216 | 2.449 | 1.817 | 56 | 3,136 | 175,616 | 7.483 | 3.826 |
| 7 | 49 | 343 | 2.646 | 1.913 | 57 | 3,249 | 185,193 | 7.550 | 3.849 |
| 8 | 64 | 512 | 2.828 | 2.000 | 58 | 3,364 | 195,112 | 7.616 | 3.871 |
| 9 | 81 | 729 | 3.000 | 2.080 | 59 | 3,481 | 205,379 | 7.681 | 3.893 |
| 10 | 100 | 1,000 | 3.162 | 2.154 | 60 | 3,600 | 216,000 | 7.746 | 3.915 |
| 11 | 121 | 1,331 | 3.317 | 2.224 | 61 | 3,721 | 226,981 | 7.810 | 3.936 |
| 12 | 144 | 1,728 | 3.464 | 2.289 | 62 | 3,844 | 238,328 | 7.874 | 3.958 |
| 13 | 169 | 2,197 | 3.606 | 2.351 | 63 | 3,969 | 250,047 | 7.937 | 3.979 |
| 14 | 196 | 2,744 | 3.742 | 2.410 | 64 | 4,096 | 262,144 | 8.000 | 4.000 |
| 15 | 225 | 3,375 | 3.873 | 2.466 | 65 | 4,225 | 274,625 | 8.062 | 4.021 |
| 16 | 256 | 4,096 | 4.000 | 2.520 | 66 | 4,356 | 287,496 | 8.124 | 4.041 |
| 17 | 289 | 4,913 | 4.123 | 2.571 | 67 | 4,489 | 300,763 | 8.185 | 4.062 |
| 18 | 324 | 5,832 | 4.243 | 2.621 | 68 | 4,624 | 314,432 | 8.246 | 4.082 |
| 19 | 361 | 6,859 | 4.359 | 2.668 | 69 | 4,761 | 328,509 | 8.307 | 4.102 |
| 20 | 400 | 8,000 | 4.472 | 2.714 | 70 | 4,900 | 343,000 | 8.367 | 4.121 |
| 21 | 441 | 9,261 | 4.583 | 2.759 | 71 | 5,041 | 357,911 | 8.426 | 4.141 |
| 22 | 484 | 10,648 | 4.690 | 2.802 | 72 | 5,184 | 373,248 | 8.485 | 4.160 |
| 23 | 529 | 12,167 | 4.796 | 2.844 | 73 | 5,329 | 389,017 | 8.544 | 4.179 |
| 24 | 576 | 13,824 | 4.899 | 2.884 | 74 | 5,476 | 405,224 | 8.602 | 4.198 |
| 25 | 625 | 15,625 | 5.000 | 2.924 | 75 | 5,625 | 421,875 | 8.660 | 4.217 |
| 26 | 676 | 17,576 | 5.099 | 2.962 | 76 | 5,776 | 438,976 | 8.718 | 4.236 |
| 27 | 729 | 19,683 | 5.196 | 3.000 | 77 | 5,929 | 456,533 | 8.775 | 4.254 |
| 28 | 784 | 21,952 | 5.292 | 3.037 | 78 | 6,084 | 474,552 | 8.832 | 4.273 |
| 29 | 841 | 24,389 | 5.385 | 3.072 | 79 | 6,241 | 493,039 | 8.888 | 4.291 |
| 30 | 900 | 27,000 | 5.477 | 3.107 | 80 | 6,400 | 512,000 | 8.944 | 4.309 |
| 31 | 961 | 29,791 | 5.568 | 3.141 | 81 | 6,561 | 531,441 | 9.000 | 4.327 |
| 32 | 1,024 | 32,798 | 5.657 | 3.175 | 82 | 6,724 | 551,368 | 9.055 | 4.344 |
| 33 | 1,089 | 35,937 | 5.745 | 3.208 | 83 | 6,889 | 571,787 | 9.110 | 4.362 |
| 34 | 1,156 | 39,304 | 5.831 | 3.240 | 84 | 7,056 | 592,704 | 9.165 | 4.380 |
| 35 | 1,225 | 42,875 | 5.916 | 3.271 | 85 | 7,225 | 614,125 | 9.220 | 4.397 |
| 36 | 1,296 | 46,656 | 6.000 | 3.302 | 86 | 7,396 | 636,056 | 9.274 | 4.414 |
| 37 | 1,369 | 50,653 | 6.083 | 3.332 | 87 | 7,569 | 658,503 | 9.327 | 4.481 |
| 38 | 1,444 | 54,872 | 6.164 | 3.362 | 88 | 7,744 | 681,472 | 9.381 | 4.448 |
| 39 | 1,521 | 59,319 | 6.245 | 3.391 | 89 | 7,921 | 704,969 | 9.434 | 4.465 |
| 40 | 1,600 | 64,000 | 6.325 | 3.420 | 90 | 8,100 | 729,000 | 9.487 | 4.481 |
| 41 | 1,681 | 68,921 | 6.403 | 3.448 | 91 | 8,281 | 753,571 | 9.539 | 4.498 |
| 42 | 1,764 | 74,088 | 6.481 | 3.476 | 92 | 8,464 | 778,688 | 9.592 | 4.514 |
| 43 | 1,849 | 79,507 | 6.557 | 3.503 | 93 | 8,649 | 804,357 | 9.644 | 4.531 |
| 44 | 1,936 | 85,184 | 6.633 | 3.530 | 94 | 8,836 | 830,584 | 9.695 | 4.547 |
| 45 | 2,025 | 91,125 | 6.708 | 3.557 | 95 | 9,025 | 857,375 | 9.747 | 4.563 |
| 46 | 2,116 | 97,336 | 6.782 | 3.583 | 96 | 9,216 | 884,736 | 9.798 | 4.579 |
| 47 | 2,209 | 103,823 | 6.856 | 3.609 | 97 | 9,409 | 912,673 | 9.849 | 4.595 |
| 48 | 2,304 | 110,592 | 6.928 | 3.634 | 98 | 9,604 | 941,192 | 9.900 | 4.610 |
| 49 | 2,401 | 117,649 | 7.000 | 3.659 | 99 | 9,801 | 970,299 | 9.950 | 4.626 |
| 50 | 2,500 | 125,000 | 7.071 | 3.684 | 100 | 10,000 | 1,000,000 | 10.000 | 4.642 |

## Section 6: COMMON EMERGENCY SERVICE ABBREVIATIONS

| | | | |
|---|---|---|---|
| **@—** | at | **ggt or ggts—** | drops |
| **AIDS—** | acquired immune deficiency syndrome | **gm or g—** | gram |
| **BP—** | blood pressure | **gpm—** | gallons per minute |
| **°C—** | degrees Celsius (Centigrade) | **hazmat—** | hazardous materials |
| **cc—** | cubic centimeter | **Hg—** | mercury |
| **CDC—** | Center for Disease Control | **HIV—** | human immunodeficiency virus |
| **cfm—** | cubic feet per minute | **Hr or hr—** | hour |
| **cm—** | centimeter | **Ht—** | height |
| **CPR—** | cardiopulmonary resuscitation | **in—** | inch |
| **DNR—** | do not resuscitate | **IV—** | intravenous |
| **DW—** | distilled water | **Kcl—** | potassium chloride |
| **D/W—** | dextrose in water | **kg—** | kilogram |
| **EKG—** | electrocardiogram | **L or l** | liter (1,000 ml) |
| **EMS—** | emergency medical service | **lb—** | pound |
| **EMT—** | emergency medical technician | **LSC—** | Life Safety Code |
| **°F—** | degrees Fahrenheit | **mg—** | milligram |
| **FD—** | fire department | **min—** | minute |
| **FL—** | friction loss | **ml or mL—** | milliliter |
| **ft—** | foot | **NaCl—** | sodium chloride |
| **gal—** | gallon | **NF—** | needed flow |

| | | | |
|---|---|---|---|
| **NFA—** | National Fire Academy | **psi—** | pounds per square inch |
| **NFPA—** | National Fire Protection Association | **pt—** | pint |
| | | **SCBA—** | self-contained breathing apparatus |
| **NPO—** | nothing by mouth | **Sp gr—** | specific gravity |
| **OR—** | operating room | **T—** | temperature |
| **oz—** | ounce | **tbsp—** | tablespoon |
| **P—** | pulse | **tsp—** | teaspoon |
| **PDP—** | pump discharge pressure | **V—** | volume |
| **pH—** | measure of acidity or alkalinity | **wt—** | weight |
| **ppm—** | parts per million | | |

# Glossary

**Absolute value** — The value of a number without regard to its sign.

**Addends** — Numbers that are added during the addition process.

**Addition** — Process of finding the total value of two or more numbers.

**Amount** — The sum of the principal and interest.

**Appliance friction loss** — The reduction in pressure resulting from increased turbulence caused by the appliance.

**Area** — A measurement of the size of a surface, usually in square inches or square meters.

**Arithmetic** — The study of numbers and their use in the basic operations of addition, subtraction, multiplication, and division.

**Atmospheric pressure** — The pressure exerted by the atmosphere (body of air) on the earth.

**Average** — A number that is representative of a group of numbers.

**Base 10** — Numbering system based on groupings of ten; also known as the decimal system.

**Blood pressure** — A measurement of the force exerted by the heart against arterial walls when the heart contracts (beats) and relaxes.

**Cardiopulmonary resuscitation (CPR)** — Procedure of providing oxygen and chest compressions to a victim whose heart has stopped beating.

**Celsius (C)** — A measurement scale for temperature on which 0° is the freezing point and 100° is the boiling point; also called centigrade.

**Combination nozzle** — A nozzle designed to provide both a straight stream and a wide fog pattern.

**Common fractions** — Numbers that represent parts of a whole unit compared to the whole.

**Customary measurement system** — A standard measurement system of weights and measures in which the inch, foot, yard, quart, pound, and gallon are used as measures of consistency.

**Decimal fractions** — Numbers that represent parts of a whole unit in multiples of 10.

**Decimal point** — A dot or period used to identify the separation of whole numbers from partial numbers.

**Density** — Weight of a substance expressed in units of mass per volume.

**Digits** — The numbers 1, 2, 3, 4, 5, 6, 7, 8, 9, and 0.

**Discount** — The amount of money deducted from the cost of goods and services.

**Discount rate** — The percent that the list price is to be reduced.

**Dividend** — The number being divided in the division process.

**Division** — The mathematical procedure of dividing or reducing a given number into two or more parts.

**Divisor** — The number indicating how many times to divide the dividend during division operations.

**Electrocardiogram (EKG)** — A graphic tracing of the electrical activity of the heart.

**Equivalent fractions** — Fractions that are equal in value and different in form.

**Estimate** — Process used to find the approximate value.

**Factors** — The numbers multiplied together during the multiplication process.

**Fahrenheit (F)** — A measurement scale for temperature on which 32° is the freezing point and 212° is the boiling point.

**Fraction** — A numerical representation of a number less than the whole stated in either a common fraction form or a decimal fraction form.

**Fraction bar** — The line, either in a horizontal or diagonal position, that separates the numerator from the denominator in a common fraction.

**Gross pay** — The amount of pay earned for hours worked before deductions are taken out.

**Grouping symbols** — The symbols parentheses ( ), brackets [ ], and braces { }, which indicate the order of operations within a combined operation problem.

**Improper fractions** — Fractions that have a numerator that is equal to or larger than the denominator.

**Integers** — All positive numbers, negative numbers, and 0, . . . −3, −2, −1, 0, 1, 2, 3 . . ., with no fractional or decimal parts; sometimes called whole numbers.

**Interest** — The amount of money charged or paid by an institution to borrow or lend money.

**Like fractions** — Fractions that have the same denominator.

**List price** — The cost of goods and services before discounts are included.

**Mass** — Measurement of the downward force exerted on an object by the earth's gravity.

**Mathematics** — The study of the measurement, properties, and relationships of quantities, using numbers and symbols.

**Mean** — The arithmetic average of a set of data.

**Median** — A number that represents the midpoint in a data set.

**Metric measurement system** — A standard decimal measuring system that is based on the meter and the kilogram.

**Minuend** — The number from which another number is subtracted during the subtraction process.

**Mixed numbers** — A number consisting of both an integer and a fraction.

**Mode** — Most frequent score in a data set.

**Multiplicand** — The number being multiplied during the multiplication process.

**Multiplication** — A mathematical procedure that increases one number by the number of times according to the size of the second number.

**Multiplier** — The number indicating how many times to add the multiplicand in the multiplication process.

**Needed flow** — The estimated flow required to extinguish a fire.

**Negative integers** — The negative numbers −1, −2, −3, −4, . . . .

**Net income** — The amount of pay received for hours worked after all deductions have been taken out; also referred to as take-home pay.

**Net price** — The price of an item after the discount has been included.

**Nozzle pressure** — The designed operating pressure for a particular nozzle.

**Nozzle reaction** — The tendency of a nozzle to move in the direction opposite of water flow.

**On-board water supply** — The water carried in a tank on the apparatus.

**Percent** — A procedure to determine the portion of the whole that is 100%.

**Periods** — Groups of three digits in the base 10 numbering system.

**Positive integers** — The positive numbers 1, 2, 3, 4, . . .; sometimes referred to as counting numbers.

**Pressure** — The force applied over an area expressed in units of force per unit area.

**Principal** — The amount of money borrowed.

**Product** — The result of the multiplication process.

**Proper fractions** — Fractions that have a smaller numerator than the denominator.

**Pulse** — Pressure of the blood felt against the wall of an artery as the heart contracts or beats.

**Quotient** — The result of the division process.

**Rate** — The percent used to calculate the interest to be charged or paid for borrowing or lending money.

**Rational numbers** — Set of numbers that includes fractions and decimal fractions.

**Real numbers** — Set of numbers that includes integers, fractions, and decimal fractions.

**Reciprocal** — The process for creating a fraction where the numerator and denominator have changed position.

**Remainder** — The number left over when the divisor cannot be equally contained in the dividend.

**Rounding** — Using an approximate value of an exact number to a given place value.

**Signed operations** — The use of positive and negative numbers in arithmetic operations.

**Smooth bore nozzle** — A nozzle designed to produce a compact solid stream of water with extended reach.

**Standard form** — Numbers written using the digits 0, 1, 2, 3, 4, 5, 6, 7, 8, and 9.
**Statistics** — Branch of mathematics that deals with the collection, analysis, interpretation, and presentation of numerical data.
**Stethoscope** — An instrument for listening to internal body sounds.
**Subtraction** — Process of finding the difference between two numbers.
**Subtrahend** — The number subtracted from the minuend in the subtraction process.
**Sum** — The result of adding two or more numbers.
**Temperature** — A measurement of the balance between heat lost and heat produced by the body.
**Term** — The length of time for a loan.
**Unlike fractions** — Fractions that have different denominators.
**Volume** — The amount of space occupied by an object expressed in cubic units.
**Whole numbers** — The numbers 0, 1, 2, 3, 4, 5, 6, 7, 8, 9, 10, 11, 12, and so on.

# ANSWERS TO ODD-NUMBERED PROBLEMS

## SECTION 1   BASIC CONCEPTS

### UNIT 1   INTRODUCTION TO NUMBERS

1. three hundred fifty-six
3. twenty-six thousand four hundred ninety-eight
5. 314 medical runs
7. 104,006 square feet or $ft^2$
9. 240
11. 2,400
13. 5,000
15. 16

### UNIT 2   ADDITION OF POSITIVE INTEGERS

1. 43
3. 200
5. 36,231
7. 700 gpm
9. 290 ft
11. 12 nozzles
13. 2,550 gal
15. 52 hr

### UNIT 3   SUBTRACTION OF POSITIVE INTEGERS

1. 14
3. 308
5. 198
7. 107 bpm
9. Personnel:  $11,839
   Operations: $2,295
   Equipment:  $516
   Total:      $13,618
11. 19 inspections left
13. 454 sections of hose are still in service
15. 10,235 mi

## UNIT 4  MULTIPLICATION OF POSITIVE INTEGERS

1.

| | 0 | 1 | 2 | 3 | 4 | 5 | 6 | 7 | 8 | 9 |
|---|---|---|---|---|---|---|---|---|---|---|
| 1 | | 1 | 2 | 3 | 4 | 5 | 6 | 7 | 8 | 9 |
| 2 | | 2 | 4 | 6 | 8 | 10 | 12 | 14 | 16 | 18 |
| 3 | | 3 | 6 | 9 | 12 | 15 | 18 | 21 | 24 | 27 |
| 4 | | 4 | 8 | 12 | 16 | 20 | 24 | 28 | 32 | 36 |
| 5 | | 5 | 10 | 15 | 20 | 25 | 30 | 35 | 40 | 45 |
| 6 | | 6 | 12 | 18 | 24 | 30 | 36 | 42 | 48 | 54 |
| 7 | | 7 | 14 | 21 | 28 | 35 | 42 | 49 | 56 | 63 |
| 8 | | 8 | 16 | 24 | 32 | 40 | 48 | 56 | 64 | 72 |
| 9 | | 9 | 18 | 27 | 36 | 45 | 54 | 63 | 72 | 81 |

3. 1,537
5. 257,180
7. 90 calls
9. 612 beats
11. 144
13. 50
15. 253,500 ml/hr

## UNIT 5  DIVISION OF POSITIVE INTEGERS

1. 66 r3
3. 32
5. 179,205 r1
7. 42 sections of hose
9. 12 min
11. 500 gpm
13. 4 r49 psi
15. 400 gpm

## UNIT 6  NEGATIVE INTEGERS, PROPERTIES OF ZERO AND ONE, EXPONENTS AND SQUARE ROOTS

1. a. −73
   b. −1,449
   c. 570
   d. −2,075

3. a. −15
   b. −384
   c. 600
   d. 851

5. a. 25
   b. −549
   c. 0
   d. 0

7. Addition:
same sign, add absolute value of numbers, and attach the sign of original numbers to answer
different sign, find difference of absolute value, and attach sign of larger number to answer

Subtraction:
rephrase into an addition problem by changing the operation sign from a minus to a plus and switching the sign of the subtrahend; proceed using rules for addition

Multiplication and Division:
same sign for both numbers always results in positive answer
different sign always results in negative answer

9. 27
11. Yes, because $4 \times 4$ or $(4^2)$ = 16
13. 2°F
15. –20 psi

## UNIT 7  COMBINED OPERATIONS WITH INTEGERS

1. $(-24) \times (6 + 8) = (-24) \times 14 = -336$
   $(8 + 6) \times (-24) = 14 \times (-24) = -336$
3. $(8 \times -4) \times 5 = (-32) \times 5 = -160$
   $8 \times (-4 \times 5) = 8 \times -20 = -160$
5. $25 \times (-14 + 2) = 25 \times (-12) = -300$
   $(25 \times -14) + (25 \times 2) = (-350) + 50 = -300$
7. a. $804
   b. $966
   c. 18 hr (rounded to ones place)
9. Vendor 1: $22 per kit (best price)
   Vendor 2: $24 per kit

# SECTION 2   COMMON FRACTIONS

## UNIT 8  INTRODUCTION TO COMMON FRACTIONS

1. a. 1/2, 7/8
   b. 14/2, 15/3
   c. 6 1/5, 2 1/2
   d. 2/5, 4/5
   e. 3/8, 4/16
3. 12
5. 3/15

7. 3/4
9. 5/8
11. a. 3/1
    b. 5/1
    c. 14/1
    d. 7/1

13. 12/24
15. 3/16
17. 11 1/4
19. 19/4

## UNIT 9  ADDITION OF COMMON FRACTIONS

1. 3/4
3. 1 1/8
5. 6 3/4 hr
7. 6 31/32 in

9. 67 19/40
11. 1 1/2 cups
13. 1"
15. 12 3/4 semester hours

17. 27 1/16 ft
19. 71 79/80 min

## UNIT 10  SUBTRACTION OF COMMON FRACTIONS

1. ½
3. ¼ in
5. ⁹⁄₁₆ in
7. 2⅞ pounds
9. 5⁹⁄₁₆
11. 242¾ lb
13. 1⅛ ft
15. 9½ in
17. 20¾ ft³
19. 4⅛ gal

## UNIT 11  MULTIPLICATION OF COMMON FRACTIONS

1. ¹⁄₈₀
3. ³⁄₁₂₈
5. 4½
7. 7¹³⁄₁₆
9. 8⅓
11. a. ¹⁄₂₅ second
    b. ⅘ second
13. 86 ft
15. 1,600⅛ sq ft
17. 62½ psi
19. $300,000

## UNIT 12  DIVISION OF COMMON FRACTIONS

1. 1½
3. 6½
5. ¹⁄₈₀
7. 6½ in
9. 27⅕ yards
11. 8¹¹⁄₂₅ hours per day
13. 1½ hours
15. 40 min
17. 15¹⁄₁₂ psi
19. 103¹⁵⁄₁₆ lives

## UNIT 13  COMBINED OPERATIONS WITH COMMON FRACTIONS

1. ¹⁹⁄₂₄
3. 1⅛
5. 5
7. ¼
9. ¹⁷⁄₃₂
11. 3,493 deaths
13. EMT: $340
    Paramedic: $372
    Combined: $712
15. Career: 25¹⁄₂₅
    Volunteer: 66⁶⁄₂₅
17. 3⁷⁄₃₆ hr
19. Proposed Budget
    Total: $28,125
    Operating: $3,515⅝
    Equipment: $7,031¼
    Personnel: $14,062½
    Training: $3,515⅝

# SECTION 3  DECIMAL FRACTIONS

## UNIT 14  INTRODUCTION TO DECIMAL FRACTIONS

1. 1.75" = 1⁷⁵⁄₁₀₀" = 1 ¾"
   95.5 gpm = 95⁵⁰⁄₁₀₀ gpm = 95½ gpm
3. 4.125 hr
5. 225.25 ml
7. 3.5
9. 2,365.92

## UNIT 15  ADDITION OF DECIMAL FRACTIONS

1. 7.8
3. 744.3535
5. 15.81 feet
7. 0.00169543
9. 64.347
11. Male coupling: 5.04"
    Female coupling: 3.99"
    Combined: 9.03"

13. $5,269.16
15. 10 hr
17. Team A: 6.4 mi
    Team B: 11.6 mi
    Team C: 5.8 mi
    Combined: 23.8 mi
19. $1,872.35

## UNIT 16  SUBTRACTION OF DECIMAL FRACTIONS

1. 2.65
3. 0.59
5. 44.274
7. 100.335 gpm
9. 43.553237
11. 295.91 gpm
13. 323.48 gal
15. 14.86 ppm
17. 1,142.63 ft
19. 254.5 mi

## UNIT 17  MULTIPLICATION OF DECIMAL FRACTIONS

1. 6.30
3. 239.754
5. 96.04
7. 0.0972
9. 1.7041
11. 13.2'
13. 33,899.36 gallons; 282,796.8 pounds
15. 288.75 ft
17. $87,692.38; $438,495.62
19. Northeast: 1.75
    North Central: 1.825
    South: 2.25
    West: 1.4

## UNIT 18  DIVISION OF DECIMAL FRACTIONS

1. .24
3. 1.49
5. 5.1
7. 6.3
9. 7
11. $21.25 each
13. 15 minutes
15. 483.33 feet per engine
17. 22 sections
19. 0.43 psi

## UNIT 19  DECIMAL AND COMMON FRACTION EQUIVALENTS

1. 0.125
3. 43.6885
5. 4.156
7. 3/16
9. 3/4
11. 98 3/5 °F
13. 2,345 3/16 sq ft
15. 1 1/8 in

## UNIT 20  COMBINED OPERATIONS WITH DECIMAL FRACTIONS

1. 29.53
3. 51.25
5. 6.76125
7. 170.5
9. Cost to Fire Department: $1,175
   Profit for County Health Department: $1,045.75

# SECTION 4   PERCENT, INTEREST, AVERAGES, AND ESTIMATES

## UNIT 21   PERCENT AND PERCENTAGES

1. a. 0.25
   b. 0.03
   c. 0.75
   d. 1.2
   e. 0.005
   f. 0.085

3. a. 17/50
   b. 23/100
   c. 2/5
   d. 1¼
   e. ¾
   f. 1/20

5. $P$ = 12.5 (or 13) sprinkler heads rejected

7. $B$ = 260
9. $P$ = 155.63
11. $R$ = .0789 or 7.89%
13. $P$ = 90 lb
15. 29 firefighters will retire; 326 firefighters will remain
17. $B$ = 20 sections of hose are on Engine 8

## UNIT 22   INTEREST AND DISCOUNTS

1. a. $I$ = $432
   b. $I$ = $700
   c. $I$ = $350
3. a. $D$ = $12.51
      $NP$ = $237.74
   b. $D$ = $17
      $NP$ = $68
5. $I$ = $437.50
   $A$ = $5,437.50
7. $D$ = $449.23
   $NP$ = $3,294.36

9. a. Loan (1-year @ 8.5%)
      $I$ = $319.43
      $A$ = $4,077.43
      Loan (2-year @ 7.25%)
      $I$ = $544.91
      $A$ = $4,302.91
      Loan (3-year @ 4.75%)
      $I$ = $535.52
      $A$ = $4,293.52
      The 1-year loan @ 8.5% interest is the best deal.

   b. Cash: $3,607.68
      $4,077.43 (loan amt)
      −$3,607.68 (cash) =
      $469.75 difference

## UNIT 23   AVERAGES AND ESTIMATES

1. a. 11.75
   b. 26.6 psi
   c. 23.75%
   d. 345 gpm

3. a. 40.4 minutes a day
   b. $0.808 average daily cost
   c. $24.24 monthly cost

5. 2,335.57 mg
7. 88.75
9. First semester: 96.5% or 97% on final

# SECTION 5   MEASUREMENT

## UNIT 24   INTRODUCTION TO MEASUREMENT

1. 64 qt
3. 11.96 in

5. 5.525 m
7. 127 mm

9. 5 m

## UNIT 25  LENGTH MEASUREMENTS

1. 250 cm
3. 425 mm
5. 8,800 yd
7. 9.323 mi
9. 30.48 m
11. a. 600 in
    b. 15.24 m
    c. 16.67 yd
    d. 1,524 cm
13. Line A:  137.2 m
              7.62 cm
    Line B:  152.4 m
              6.35 cm
    Line C:  45.7 m
              4.45 cm
    Line D:  30.48 m
              3.81 cm
    Line E:  6.096 m
              10.16 cm

## UNIT 26  AREA AND PRESSURE MEASUREMENTS

1. 16 m$^2$
3. 50.27 cm$^2$
5. 95.03 in$^2$
7. 3,375 ft$^2$
9. 11.94 psi
11. $P = 2.5$ lb/ft$^2$
13. $A = 107.8$ ft$^2$
15. 28 yd$^2$

## UNIT 27  SOLID AND FLUID VOLUME MEASUREMENTS

1. 125 m$^3$
3. 464.1 cm$^3$
5. 12.7 gal
7. 1,728 in$^3$
9. 77,799 gal
11. 4.94 l
13. 28.13 gal
15. 1,321 gal

## UNIT 28  MASS AND DENSITY MEASUREMENTS

1. 5,143,824 g
3. 1.102 lb
5. 41,700 lb
7. 3,120 lb
9. 792.3 lb per min

## UNIT 29  TEMPERATURE MEASUREMENTS

1. 15°C
3. 4°C
5. 28°C
7. 50°F
9. 118°F
11. 98.6°F
13. 1,220.7°F
15. –43°C

# SECTION 6  STATISTICS, CHARTS, AND GRAPHS

## UNIT 30  INTRODUCTION TO STATISTICS

1. Mode = 8
   Median = 10
   Mean = 12.4

3. 
| | FF1 | FF2 | FF3 |
|---|---|---|---|
| Mode | 1 mi | 1 mi | 1 mi |
| Median | 1 mi | 1 mi | 1 mi |
| Mean | 1.2 mi | 0.96 mi | 1.4 mi |

5. Mode = 3 min
   Median = 4.5 min
   Mean = 4.46 min

## Unit 31 Line Graphs

1. a. 20 each year (40 total)
   b. 1996 (30)
   c. 1994 (18)
   d. rising
3. a. y-axis: number of runs
      x-axis: quarter
   b. Ambulance 1
   c. Ambulance 1: 2nd quarter
      Ambulance 2: 2nd quarter
      Ambulance 3: 4th quarter
   d. Trends for each ambulance appear parallel: from 1st to 2nd quarter runs slow down, especially for ambulances 2 and 3; from the 2nd to 3rd quarter runs increase for each ambulance; finally, from 3rd to 4th quarter runs decrease for each ambulance.
5. 4 seconds:   22 ft/sec
   6 seconds:   30 ft/sec
   11 seconds:  50 ft/sec
   16 seconds:  70 ft/sec

## Unit 32 Bar Graphs

1. a. prunes (10.9 g), bran cereal (8.5 g), beans (6.5 g)
   b. broccoli (0.6 g), carrot (1.6 g), corn & peanuts (2.1 g each)
   c. 5 types (bran cereal, apple, beans, prunes, raspberries)
   d. corn and peanuts
3. a. 1983 (13 deaths), 1985 (11 deaths), 1993 (10 deaths)
   b. 1987 (5 deaths), 1991 (4 deaths), 1993 (10 deaths)
   c. 1992 (2 deaths), 1991 (4 deaths), 1988 (4 deaths)
   d. 1983 to 1992: strong decline in deaths associated with fireworks accidents
      1983 to 1993: strong decline until 1993, when there was a sharp rise
5.

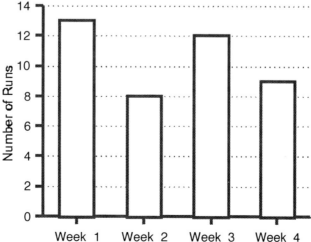

*200   Answers to Odd-Numbered Problems*

## UNIT 33   PIE CHARTS

1.  a. home injuries (29.1%) and auto accidents (26.1%)
    b. respiratory problems (8.8%)
    c. 14.0%
    d. 11.8%
3.  a.
5.  a. hand/wrist (24%) and eyeball (24%)
    b. hand/wrist (24%), eyeball (24%), and face (15%)
    
    b. stress (50%) and struck by contact with object (32.6%)
    
    c. arm/elbow (3%)
    d. 48% (eyeball 24% + hand/wrist 24% = 48%)

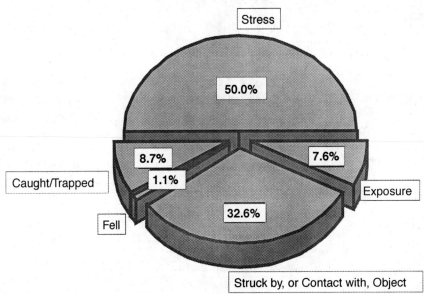

**Firefighter Deaths by Cause of Injury, 1996**

*NFPA Journal, July/August 1997*

# SECTION 7   FORMULAS AND EQUATIONS

## UNIT 34   INTRODUCTION TO FORMULAS AND EQUATIONS

1. $r = \frac{1}{2}d$ or $r = \frac{d}{2}$
   where   $r$ = radius
   $d$ = diameter

3. $V = L \times W \times H$
   where   $V$ = volume
   $L$ = length
   $W$ = width
   $H$ = height
   Note: units must be the same

5. $NF = \dfrac{A}{3 \text{ ft}^2/\text{gpm}}$
   where   $NF$ = needed flow
   $A$ = area in square feet (ft$^2$)
   3ft$^2$/gpm = constant

7. $P = \dfrac{F}{A}$
   where   $P$ = pressure in pounds per square inch (lb/in$^2$ or psi) or pounds per square foot
   $F$ = force in pounds (lb)
   $A$ = surface area in square inches (in$^2$) or square feet (ft$^2$)

9. $PDP = FL + NP + EL + AFL$
   where   $PDP$ = pump discharge pressure
   $FL$ = friction loss
   $NP$ = nozzle pressure
   $EL$ = pressure gain or loss due to elevation
   $AFL$ = appliance friction loss

## UNIT 35   COMMON EMERGENCY SERVICE FORMULAS AND EQUATIONS

1. a. $NR$ = 11.04 lb
   b. $NR$ = 47.98 lb or 48.0 16
3. $FR$ = 41.7 gtt/min
5. a. 56.3 psi
   b. 36.3 psi
7. $h$ = 100.0 ft
9. 3.3 mg